Pre-GED Connection™

Science

by Marion Castellucci

LiteracyLink® is a joint project of PBS,
Kentucky Educational Television,
the National Center on Adult Literacy,
and the Kentucky Department of Education.

This project is funded in whole,
or in part, by the Star Schools Program
of the USDE under contract #R203D60001.

Acknowledgments

LiteracyLink® Advisory Board
Lynn Allen, Idaho Public Television
Anthony Buttino, WNED-TV
Anthony Carnevale, Educational
Testing Service
Andy Chaves, Marriott International, Inc.
Patricia Edwards, Michigan State University
Phyllis Eisen, Center for Workforce Success National
Association of Manufacturers
Maggi Gaines, Baltimore Reads, Inc.
Marshall Goldberg, Association of Joint Labor
Management Educational Programs
Milton Goldberg, National Alliance
for Business
Neal Johnson, Association of Governing Boards of
Universities and Colleges
Cynthia Johnston, Central Piedmont Community
College
Sandra Kestner, Kentucky Department for Adult
Education and Literacy
Thomas Kinney, American Association of Adult and
Continuing Education
Dale Lipschultz, American Library Association
Lennox McLendon, National Adult Education
Professional Development Consortium
Cam Messina, KLRN
Patricia Miller, KNPB
Cathy Powers, WLRN
Ray Ramirez, U.S. Department of Education
Emma Rhodes, (retired) Arkansas Department of
Education
Cynthia Ruiz, KCET
Tony Sarmiento, Worker Centered Learning,
Working for America Institute
Steve Steurer, Correctional
Education Association
LaShell Stevens-Staley, Iowa PTV
Fran Tracy-Mumford, Delaware Department of
Adult/Community Education
Terilyn Turner, Community Education,
St. Paul Public Schools

LiteracyLink®
Ex Officio Advisory Board
Joan Auchter, GED Testing Service
Barbara Derwart, U.S. Department of Labor
Cheryl Garnette, OERI, U.S.
Department of Education
Andrew Hartman, National Institute
for Literacy
Mary Lovell, OVAE, U.S. Department
of Education
Ronald Pugsley, OVAE, U.S. Department
of Education
Linda Roberts, U.S. Department of Education
Joe Wilkes, OERI, U.S. Department of Education

LiteracyLink® Partners
LiteracyLink® is a joint project of:
Public Broadcasting Service,
Kentucky Educational Television,
National Center on Adult Literacy, and the
Kentucky Department of Education.

Content Design and Workbook
Editorial Development
Learning Unlimited, Oak Park, Illinois
Design and Layout
By Design, Lexington, Kentucky
Project Coordinators
Milli Fazey, KET, Lexington, Kentucky
Margaret Norman, KET, Lexington, Kentucky

This project is funded in whole, or in part, by the
Star Schools Program of the USDE under contract
#R203D60001.

PBS LiteracyLink® is a registered mark of the
Public Broadcasting Service.

The editor has made every effort to trace the ownership of
all copyrighted material, and necessary permissions have
been secured in most cases. Should there prove to be any
question regarding the use of any material, regret is here
expressed for such error. Upon notification of any such
oversight, proper acknowledgment will be made in future
editions.

Printed in the United States of America.
ISBN # 1-881020-48-7
ISBN # 978-1-881020-48-6

Contents

Introduction..4

Science Pretest...6

Science Pretest Answers
    and Explanations 16

Program 16 Life Science

Before You Watch.....................................18

After You Watch 19

Cells—The Building Blocks of Life 20

Human Body Systems 24

Ecosystems and the Environment..............28

Comprehend Science Materials.................32

Understand Diagrams 34

GED Review: Life Science 36

Program 17 Earth and Space Science

Before You Watch......................................38

After You Watch 39

The Changing Earth..................................40

People and the Environment......................44

The Solar System and the Universe 48

Apply Science Information.......................52

Understand Graphs 54

GED Review:
    Earth and Space Science 56

Program 18 Chemistry

Before You Watch......................................58

After You Watch 59

Matter: What Everything Is
    Made Of...60

Elements, Compounds,
    and Mixtures...64

Chemical Reactions 68

Analyze Science Information 72

Understand Charts and Tables...................74

GED Review: Chemistry 76

Program 19 Physics

Before You Watch......................................78

After You Watch 79

Motion, Work, and Energy 80

Waves..84

Magnetism and Electricity........................88

Evaluate Science Information....................92

Understand Diagrams 94

GED Review: Physics 96

Science Posttest 98

Science Posttest Answers
    and Explanations 108

Answers and Explanations.......................110

Science Resources 122

Introduction

Welcome to *Pre-GED Science*. This workbook is part of the *LiteracyLink®* multimedia educational system for adult learners and educators. The system includes *Pre-GED Connection*, which builds a foundation for GED-level study and *GED Connection*, which learners use to study for the GED Tests. *LiteracyLink* also includes *Workplace Essential Skills*, which targets upgrading the knowledge and skills needed to succeed in the world of work.

Pre-GED CONNECTION
consists of these educational tools:

26 VIDEO PROGRAMS shown on public television and in adult learning centers

ONLINE MATERIALS available on the Internet at http://www.pbs.org/literacy

FIVE Pre-GED COMPANION WORKBOOKS
Language Arts, Writing
Language Arts, Reading
Social Studies
Science
Mathematics

Instructional Programs

Pre-GED Connection consists of 26 instructional video programs and five companion workbooks. Each *Pre-GED Connection* workbook lesson accompanies a video program. For example, the first lesson in this book is *Program 16—Life Science*. This workbook lesson should be used with *Pre-GED Connection Video Program 16—Life Science*. In addition, you can go online to www.pbs.org/literacy and click the *Pre-GED Science* link.

Who's Responsible for LiteracyLink®?

LiteracyLink was developed through a five-year grant by the U.S. Department of Education. The following partners have contributed to the development of the *LiteracyLink* system:

PBS Adult Learning Service

Kentucky Educational Television (KET)

The National Center on Adult Literacy (NCAL) of the University of Pennsylvania

The Kentucky Department of Education

All of the *LiteracyLink* partners wish you the very best in meeting all of your educational goals.

Getting Started with *Pre-GED Connection Science*

Before you start using the workbook, take some time to preview its features.

1. Take the **Pretest** on page 6. This will help you decide which areas you need to focus on. You should use the evaluation chart on page 17 to develop your study plan.

2. Work through the **workbook lessons**—each one corresponds to a video program.

    The *Before You Watch* feature sets up the video program:
      - **Think About the Topic** gives a brief overview of the video
      - **Prepare to Watch the Video** is a short activity with instant feedback that shows how everyday knowledge can help you better understand the topic
      - **Lesson Goals** highlight the main ideas of each video and workbook lesson
      - **Terms** introduces key science vocabulary

    The *After You Watch* feature helps you evaluate what you have just seen in the program:

      - **Think About the Program** presents questions that focus on key points from the video
      - **Make the Connection** applies what you have learned to real-life situations

    Three *Science Skills* sections correspond to key concepts in the video program.

    The *Thinking Skill* sections prepare you for the types of critical thinking questions that you will see on the GED.

    The *Graphic Skill* sections introduce you to the charts, diagrams, and graphs that you will see on the GED.

    GED Practice allows you to practice with the types of problems that you will see on the actual test.

3. Take the **Posttest** on page 98 to determine your progress and whether you are ready for GED-level work.

4. Use the **Answer Key** to check your answers.

5. Refer to the **Science Resources** at the back of the book as needed.

For Teachers

Portions of *LiteracyLink* have been developed for adult educators and service providers. Teachers can use Pre-GED lesson plans in the *LiteracyLink Teacher's Guide* binder. This binder also contains lesson plans for *GED Connection* and *Workplace Essential Skills*.

Science Pretest

The Science Pretest on the following pages is similar to the GED Science Test. However, it has only 25 items, compared to 50 items on the actual GED Science Test.

This pretest consists of short passages, charts, tables, diagrams, and graphs. Each passage or graphic is followed by one or more multiple-choice questions. Read each passage, study the graphics, and then answer the questions. You may refer back to the passage or graphic whenever you wish.

The purpose of the Pretest is to evaluate your science knowledge and thinking skills. Do not worry if you cannot answer every question or if you get some questions wrong. The Pretest will help you identify the content areas and skills that you need to work on.

Directions

1. Read the sample passage and test item on page 7 to become familiar with the test format.

2. Take the test on pages 8 through 15. Read each passage, study the graphics, if any, and then choose the best answer to each question.

3. Record your answers on the answer sheet below, using a No. 2 pencil.

4. Check your work against the Answers and Explanations on page 16.

5. Enter your scores in the evaluation charts on page 17.

SCIENCE PRETEST ■ ANSWER SHEET

Name _____   Date _____

Class _____

1. ①②③④⑤	6. ①②③④⑤	11. ①②③④⑤	16. ①②③④⑤	21. ①②③④⑤
2. ①②③④⑤	7. ①②③④⑤	12. ①②③④⑤	17. ①②③④⑤	22. ①②③④⑤
3. ①②③④⑤	8. ①②③④⑤	13. ①②③④⑤	18. ①②③④⑤	23. ①②③④⑤
4. ①②③④⑤	9. ①②③④⑤	14. ①②③④⑤	19. ①②③④⑤	24. ①②③④⑤
5. ①②③④⑤	10. ①②③④⑤	15. ①②③④⑤	20. ①②③④⑤	25. ①②③④⑤

Sample Passage and Test Item

The following passage and test item are similar to those you will find on the Science Pretest. Read the passage and the test item. Then go over the answer sheet sample and explanation of why the correct answer is correct.

Question 0 refers to the following passage.

Substances that may be present in groundwater and surface water include:

- **Pathogens (protists, bacteria, and viruses)** may come from sewage, septic systems, or wildlife and their wastes.
- **Inorganic substances (salts and metals)** may be natural or from industrial and farm waste.
- **Pesticides and herbicides** come from farms, businesses, and home use.
- **Organic chemicals** result from industrial processes and oil production. They may also come from gas stations and septic systems.
- **Radioactive contaminants** may occur naturally or may come from oil production or mining.

0. According to the passage, which substances in groundwater may come from wildlife?

  (1) pathogens   **(4)** organic chemicals

  (2) inorganic substances   **(5)** radioactive contaminants

  (3) pesticides and herbicides

Marking the Answer Sheet

0. ①②③④⑤

The correct answer is **(1) pathogens.** Therefore, answer space 1 is marked on the answer sheet, as shown above. The space should be filled in completely using a No. 2 pencil. If you change your mind about an answer, erase it completely.

Answer and Explanation

(1) pathogens (Comprehension) According to the list, wildlife and their wastes are one of the sources of pathogens in the water supply. To find this answer, you had to skim the list of substances to see which one comes from wildlife. None of the other answer choices mentions wildlife.

Science Pretest

Choose the <u>one best answer</u> to the questions below.

<u>Questions 1 and 2</u> refer to the following passage.

Every year people get sick from eating hamburger tainted with *E. coli* bacteria. Clean beef can be contaminated at meat-processing plants. The bacteria live in waste on the animals' hides and in their intestines. When meat is ground, the *E. coli* can be accidentally mixed in. If *E. coli* stays in the grinding equipment, it can pass into thousands of pounds of ground beef.

Cooking hamburgers well done (to an internal temperature of 160°F) will kill *E. coli*. However, many people prefer their hamburgers rare—and risky.

1. Which process is often the source of *E. coli* contamination of hamburger?
   (1) transporting cattle to processing plants
   (2) butchering cattle at processing plants
   (3) grinding beef at processing plants
   (4) adding seasoning to hamburger meat
   (5) cooking hamburgers well done

2. Beth always cooks well-done hamburgers for her family. If she wanted to further reduce her family's chances of *E. coli* contamination, what else might she do?
   (1) buy hamburger patties
   (2) buy ground beef
   (3) mix ground beef with seasonings
   (4) grind the beef herself
   (5) cook the hamburgers on a grill

3. Some animals that migrate yearly, including some birds and fish, have magnetic particles in their brain cells. Humans also have magnetic particles in their brain cells, but the concentration of magnetic particles is lower in human brain cells than in the brain cells of migrating animals.

   What is the best explanation for this difference?
   (1) Humans are bigger.
   (2) Humans have larger brains.
   (3) Humans live longer.
   (4) Humans walk on two legs.
   (5) Humans don't migrate annually.

4. When a nonnative species enters a new ecosystem, it can drive out native species. For example, zebra mussels are native to the Caspian and Black Seas. When they entered the Great Lakes, they spread uncontrollably because they have no natural enemies there.

   Species enter new waters when ships pump out ballast water containing organisms from the last port the ship visited. In 1996, a law was passed requiring ocean ships to empty and refill their ballast tanks mid-ocean.

   Which of the following would be evidence that the 1996 law is working?
   (1) a decrease in nonnative species
   (2) a decrease in native species
   (3) an increase in nonnative species
   (4) an increase in zebra mussels
   (5) a decrease in zebra mussels

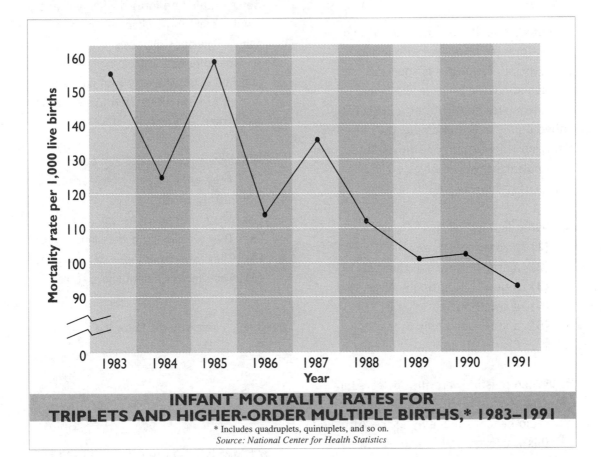

INFANT MORTALITY RATES FOR TRIPLETS AND HIGHER-ORDER MULTIPLE BIRTHS,* 1983–1991

* Includes quadruplets, quintuplets, and so on.
Source: National Center for Health Statistics

5. Between 1987 and 1991, the infant mortality rate for triplets and higher-order multiple births
   (1) rose sharply
   (2) first rose and then fell sharply
   (3) first fell and then rose sharply
   (4) generally declined
   (5) remained the same

6. The data on this graph would be of the most professional interest to
   (1) a baby-equipment manufacturer
   (2) a preschool teacher
   (3) a nurse who works in an intensive-care unit for newborns
   (4) a physician who works with patients having heart disease
   (5) a nutritionist

7. Which of the following statements is supported by data from the graph?
   (1) Overall infant mortality rates rose between 1983 and 1991.
   (2) Infant mortality rates for triplets and higher order multiple births are much higher than those for babies born singly.
   (3) There was a peak in the infant mortality rate for triplets and higher-order multiple births in 1984.
   (4) The greatest one-year decline in infant mortality rates for triplets and higher-order multiple births took place between 1985 and 1986.
   (5) The high mortality rate for triplets and higher-order multiple births is related to the low birth weight of these infants.

Questions 8 and 9 refer to the following passage.

A neutron star is extremely dense, with a mass greater than that of our sun. However, all its matter is squeezed into a space about 12 miles in diameter. A neutron star that rotates quickly is called a pulsar. For example, the pulsar in the Crab Nebula spins 30 times per second. As a pulsar rotates, it gives off regular bursts of radio waves.

8. What is a pulsar?
   (1) a rotating nebula
   (2) a star larger than the sun
   (3) a quickly rotating neutron star
   (4) a small planet
   (5) a star found only in the Crab Nebula

9. When pulsars were first detected in 1967, some people thought they were radio beacons made by aliens. Although this was eventually ruled out, which of the following facts could have supported the theory that pulsars were made by alien civilizations?
   (1) Pulsars are extremely dense.
   (2) Pulsars have a mass greater than that of our sun.
   (3) Pulsars have a diameter of about 12 miles.
   (4) The Crab Nebula pulsar rotates about 30 times per second.
   (5) Pulsars give off regular bursts of radio waves.

10. A crater is a circular depression in rock. Some craters are formed when volcanoes erupt. Others form when underground salt or limestone dissolves and the surface collapses. Retreating glaciers and falling groundwater also can cause craters to form. Others form when large meteorites hit Earth's surface.

    What do all craters have in common?
    (1) They are caused by meteorites hitting Earth.
    (2) They are caused by volcanoes.
    (3) They are circular depressions in rock.
    (4) They form from loose soil and sand.
    (5) They form when glaciers retreat.

Question 11 refers to the following diagram.

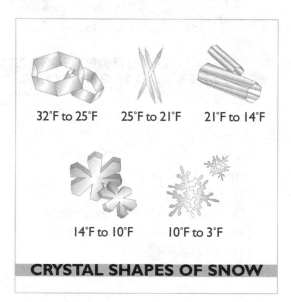

32°F to 25°F   25°F to 21°F   21°F to 14°F

14°F to 10°F   10°F to 3°F

CRYSTAL SHAPES OF SNOW

11. What type of change occurs in snowflakes as the temperature drops?
    (1) They become larger.
    (2) They become smaller.
    (3) Their shapes become simpler.
    (4) Their shapes become rounder.
    (5) Their shapes become more complex.

Code	Material	Examples
1 PETE	Polyethylene terephthalate	Soft-drink bottles, peanut butter jars
2 HDPE	High-density polyethylene	Milk jugs, water jugs
3 V	Vinyl/polyvinyl chloride	Shampoo bottles
4 LDPE	Low-density polyethylene	Ketchup bottles
5 PP	Polypropylene	Squeeze bottles, food storage containers
6 PS	Polystyrene	Fast food packaging

RECYCLING CODES FOR PLASTIC CONTAINERS

Source: Society of the Plastics Industry

12. What is the most likely reason that the Society of the Plastics Industry developed these codes?
- **(1)** to rank plastics from most to least harmful to the environment
- **(2)** to guide plastics manufacturers in using the right materials
- **(3)** to give consumers a guide to the uses and costs of plastics
- **(4)** to help recyclers separate different plastics for recycling
- **(5)** to encourage the disposal of plastics in landfills

13. Carmella's city recycles only plastics coded 1 and 2. According to the table, which of the following items should Carmella put in her recycle bin?
- **(1)** food storage containers
- **(2)** milk jugs
- **(3)** fast-food packaging
- **(4)** ketchup bottles
- **(5)** shampoo bottles

Questions 14 and 15 refer to the following passage.

The *Monitor,* an ironclad ship, sank during the Civil War. Recently, parts of the ship were raised from the sea bottom. After more than 100 years in water, the iron was weak and brittle. Thus, raising parts to the surface was a delicate task. Once the iron parts reached the air, iron chloride crystals could form on them, weakening them further.

To prevent this, the parts were rushed to a museum, where they were placed in vats of water and sodium carbonate or sodium hydroxide. These solutions conduct electricity. Using the process of electrolysis, scientists ran electricity through the solution. This broke up the iron chloride crystals and gently loosened any crusty material.

14. The iron parts were lifted carefully to
  (1) avoid disturbing marine animals
  (2) avoid straining the equipment
  (3) prevent the iron from breaking
  (4) stop iron chloride crystals from forming
  (5) avoid loosening crusty material

15. Based on the definition in the passage, which of the following would be another use of electrolysis?
  (1) electroplating—electricity passes through a liquid, causing an object in the liquid to gain a metal coating
  (2) etching—an acid is used to eat away metal
  (3) cracking—complex chemicals containing carbon are broken down
  (4) roasting—ores of metal are heated with air to purify them
  (5) smelting—an ore is melted to remove impurities

16. Spontaneous combustion sometimes occurs when materials are stored in large quantities. Heat builds up in the material because of oxidation, a reaction involving oxygen or hydrogen. Because the material is packed tightly, the heat cannot escape into the surrounding air. Instead, it increases until the ignition point of the material is reached. Then the material bursts into flames.

Which of the following is an example of spontaneous combustion?
  (1) grain catching fire in a silo
  (2) charcoal burning on a grill
  (3) gas burning on a stovetop
  (4) candles burning on a table
  (5) forest fires starting from lightning

17. Approximately 92 elements occur naturally on Earth, and scientists have made about two dozen more in lab experiments. The latest, element number 118, was made in Berkeley, California, in the late 1990s. In 2002, the Berkeley lab gave up its claim to element 118. No one had been able to repeat the manufacture of this element.

Why did the Berkeley lab give up its claim to element 118?
  (1) Only elements up to number 92 occur in nature.
  (2) The results of a scientific experiment must be repeatable.
  (3) Element 118 was a compound rather than an element.
  (4) The element had been created earlier in another lab.
  (5) Another lab was able to repeat the manufacture of element 118.

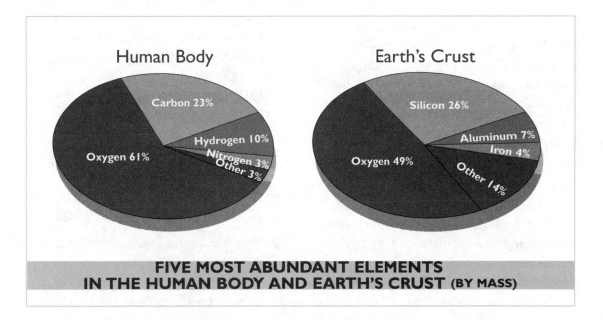

Human Body

Carbon 23%

Hydrogen 10%

Oxygen 61%

Nitrogen 3%
Other 3%

Earth's Crust

Silicon 26%

Aluminum 7%

Iron 4%

Oxygen 49%

Other 14%

**FIVE MOST ABUNDANT ELEMENTS
IN THE HUMAN BODY AND EARTH'S CRUST (BY MASS)**

18. How are the composition of the human body and the composition of Earth's crust similar?

(1) Both contain mostly carbon.

(2) Both are about one-quarter silicon.

(3) Oxygen is the most abundant element in both.

(4) Trace amounts of aluminum are found in both.

(5) Both are about one-tenth hydrogen.

19. Which data from the graphs support the conclusion that the human body is made mostly of water (H_2O) by mass?

(1) Hydrogen and oxygen make up about 71% of the mass of the human body.

(2) Carbon and oxygen make up about 84% of the mass of the human body.

(3) Nitrogen is the fourth most abundant element in the human body.

(4) Hydrogen makes up about 10% of the human body.

(5) Other elements make up about 3% of the human body.

Questions 20 and 21 refer to the following passage.

Surgeons usually push a stiff needle through the brain to reach a tumor. Unfortunately, this process can damage sensitive areas in the path of the needle. A new system uses magnetism to guide surgical instruments along the path least likely to do damage. In this system, a tiny magnet attached to a wire is put inside a catheter—a thin, flexible tube. The surgeon sits at a computer and directs the movement of the magnet in any direction. Electronic impulses made by the computer create a sequence of magnetic fields around the patient's head. These magnetic fields guide the magnet and catheter to the right location. Once the magnet reaches the tumor, the magnet is gently pulled out of the catheter, and a surgical tool is inserted in its place.

20. What is the main advantage of the magnet system as compared with the needle system?
   (1) The magnet can attract a tumor more effectively than a needle can.
   (2) The magnet can attract the surgical instrument, and a needle cannot.
   (3) The magnet can locate a tumor better than a needle can.
   (4) The magnet can be left inside the brain, and a needle cannot.
   (5) The magnet can be directed more precisely than a needle can.

21. Which of the following conclusions about magnetism is supported by the passage?
   (1) Magnetism can cure cancer.
   (2) The magnetic fields pass through the patient's skull.
   (3) Magnetic fields cut out the tumor.
   (4) Magnets have a north pole and a south pole.
   (5) The magnet attracts the surgical tool to the correct location.

22. Most substances expand when heated, although at different rates. Engineers designing machines or structures must account for the rates at which materials expand when heated.

Which of the following uses this principle?
   (1) roads with expansion joints between concrete slabs
   (2) water barrels with metal braces for strengthening
   (3) concrete reinforced with iron rebar
   (4) coins made of alloys of several metals
   (5) copper wire insulated with plastic

23. A child is shown a large pillow and a small book, each with the same mass. When asked which is heavier, he says, "The pillow." What is the child's misconception?
   (1) He thinks size and mass are unrelated.
   (2) He thinks large size means large mass.
   (3) He thinks the pillow is very big.
   (4) He thinks that mass increases as size decreases.
   (5) He thinks that mass decreases as size increases.

<u>Questions 24 and 25</u> refer to the following information and diagram.

A police officer parked by the side of a road can use a radar gun to identify speeders. The fact that he is stationary and the cars are moving affects the reading on the radar gun. Nearby cars are hit by the radar waves at a greater angle than cars that are far off, as shown in the diagram below. This reduces the frequency of the waves returning to the radar gun. As a result, the displayed speed of the nearby car is several miles per hour less than its actual speed. The closer the speeding car, the less accurate the radar gun.

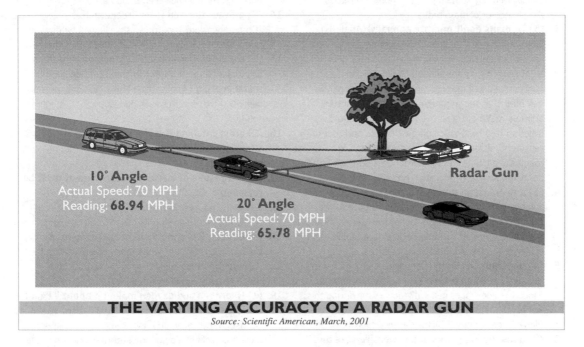

10° Angle
Actual Speed: 70 MPH
Reading: **68.94** MPH

20° Angle
Actual Speed: 70 MPH
Reading: **65.78** MPH

Radar Gun

THE VARYING ACCURACY OF A RADAR GUN

Source: Scientific American, March, 2001

24. The radar gun reading and the actual speed of a car differ. This suggests that stationary police officers should
   (1) judge by eye the speed of passing cars
   (2) expect cars to slow down as they pass
   (3) learn how to estimate the actual speed from the gun speed
   (4) take radar gun readings only at night
   (5) ticket only cars going 20 or more miles per hour above the speed limit

25. What should police officers do to ensure they get the most accurate reading possible from a radar gun?

They should
   (1) park at a right angle to the road and target the nearest cars
   (2) park far from the road and target the most distant cars
   (3) park far from the road and target the nearest cars
   (4) park close to the road and target the most distant cars
   (5) park close to the road and target the nearest cars

Answers and explanations begin on page 16.

Science Pretest Answers and Explanations

1. **(3) grinding beef at processing plants**
(Comprehension) According to the first paragraph, *E. coli* can be accidentally mixed in when beef is ground. It also can remain in the grinders, contaminating even more beef.

2. **(4) grind the beef herself** (Application) Since grinding at a meat plant is often the source of *E. coli* tainting, grinding beef at home decreases the chance of contamination.

3. **(5) Humans don't migrate annually.** (Analysis) Both humans and animals that migrate annually have magnetic particles in their brain cells. However, animals that migrate annually have a higher concentration of magnetic particles. Therefore, you can infer that the magnetic particles in migrating animals may have a role in navigation, similar in function to a magnetic compass needle, and that the reason humans have a lower concentration of magnetic particles in their brain cells is because they don't follow yearly migration patterns.

4. **(1) a decrease in nonnative species** (Evaluation) If ships are changing ballast water in mid-ocean rather than in port, the number of nonnative species they introduce into ports should decline. That would be evidence the law is effective.

5. **(4) generally declined** (Comprehension) First locate the years 1987 through 1991 on the horizontal axis. Since the line slopes generally downward during this period, the infant mortality rate for triplets and higher-order multiple births declined overall.

6. **(3) a nurse who works in an intensive-care unit for newborns** (Application) Such a nurse is likely to treat newborn triplets and would be interested in statistics relating to the infant death rate for his or her patients.

7. **(4) The greatest one-year decline in infant mortality rates for triplets and higher-order multiple births took place between 1985 and 1986.** (Evaluation) The steepest downward slope of the trend line occurs between 1985 and 1986, with the infant mortality rate falling from approximately 160 to 113 deaths per 1,000 live births.

8. **(3) a quickly rotating neutron star** (Comprehension) A pulsar is defined in the third sentence of the passage.

9. **(5) Pulsars give off regular bursts of radio waves.** (Evaluation) The regularity of the radio waves suggested that intelligent beings could have been transmitting them.

10. **(3) They are circular depressions in rock.** (Comprehension) According to the passage, craters form from different causes, but they all are circular depressions in rock.

11. **(5) Their shapes become more complex.** (Analysis) As the temperature drops, the snow crystal shapes generally become more complicated, with the most complex shapes forming at 10°F to 3°F.

12. **(4) to help recyclers separate different plastics for recycling** (Analysis) Having a simple code to identify plastics would be most useful to people who need a quick and easy way to identify the type of plastic a container is made of. The coding system in the chart was developed to help consumers and professional recyclers sort plastic to be reprocessed. The code is not helpful in achieving any of the other purposes listed as choices.

13. **(2) milk jugs** (Application) Of the items listed, only milk jugs have a code 1 or a code 2.

14. **(3) prevent the iron from breaking** (Comprehension) The iron was weakened by its time in the water, so lifting it out had to be done carefully.

15. **(1) electroplating—electricity passes through a liquid, causing an object in the liquid to gain a metal coating** (Application) This is the only process described that involves running electricity through a solution.

16. **(1) grain catching fire in a silo** (Application) Densely packed grain can heat up and eventually reach ignition.

17. **(2) The results of a scientific experiment must be repeatable.** (Analysis) Even the lab that had originally made element 118 failed to make it again, so the lab had to give up its claim to the element.

18. **(3) Oxygen is the most abundant element in both.** (Analysis) The graphs show that oxygen is the most common single element in the human body (61%) and in Earth's crust (49%).

19. **(1) Hydrogen and oxygen make up about 71% of the mass of the human body.** (Evaluation) Water consists of hydrogen (H) and oxygen (O). Together these elements make up almost three-quarters of the body's mass.

20. **(5) The magnet can be directed more precisely than a needle can.** (Analysis) The needle is pushed in a straight line, but the magnet can be guided (via the surgeon using a computer) along any path.

21. **(2) The magnetic fields pass through the patient's skull.** (Evaluation) The passage states that the computer uses electronic impulses to generate magnetic fields outside the patient's head. These fields direct the magnet, which is inside the patient's head. The fact that the magnet can be moved means that the skull is not a barrier to the magnetic fields.

22. **(1) roads with expansion joints between concrete slabs** (Application) This is the only example in which the design of the material or structure is related to how the structure responds to heating and cooling.

23. **(2) He thinks large size means large mass.** (Evaluation) It's a common mistake to think that a large object must have a large mass (and therefore weigh a lot). Since densities vary, this is not so.

24. **(3) learn how to estimate the actual speed from the gun speed** (Comprehension) In fact, officers are taught how to do this.

25. **(4) park close to the road and target the most distant cars** (Analysis) This has the effect of reducing the angle at which the radar waves hit the moving car. Parking far from the road and targeting the nearest cars both increase the angle of the radar waves.

Evaluation Charts for Science Pretest

Follow these steps for the most effective use of this chart:

- Check your answers against the Answers and Explanations on page 16.
- Use the following charts to circle the questions you answered correctly.
- Total your correct answers in each row (across) for science subject areas and each column (down) for thinking skills.

You can use the results to determine which subjects and graphics skills you need to focus on.

- The column on the left of the table indicates the KET Pre-GED video program and its corresponding lesson in this workbook.
- The column headings—*Comprehension, Application, Analysis,* and *Evaluation*—refer to the type of thinking skills needed to answer the questions.

SUBJECT AREAS AND THINKING SKILLS					
Program	Comprehension (pp. 32–33)	Application (pp. 52–53)	Analysis (pp. 72–73)	Evaluation (pp. 92–93)	Total for Science Subjects
16 Life Science (pp. 18–37)	1, 5	2, 6	3	4, 7	____/7
17 Earth and Space Science (pp. 38–57)	8, 10	13	11, 12	9	____/6
18 Chemistry (pp. 58–77)	14	15, 16	17, 18	19	____/6
19 Physics (pp. 78–97)	24	22	20, 25	21, 23	____/6
Total for Skills	____/6	____/6	____/7	____/6	

Many of the questions on the GED Science Test are based on charts, tables, diagrams, and graphs.

- Use the chart below to circle the graphics-based questions that you answered correctly.
- Identify your strengths and weaknesses in interpreting graphics by counting the number of questions you got correct for each type of graphic.

GRAPHIC SKILLS			
Diagrams (pp. 34–35, 94–95)	Charts and Tables (pp. 74–75)	Graphs (pp. 54–55)	Total for Graphics
11, 24, 25	12, 13	5, 6, 7, 18, 19	____/10

Life Science

LESSON GOALS

SCIENCE SKILLS

- Understand cell structures and functions
- Learn how the human body works as a system
- Analyze people's impact on the natural environment

THINKING SKILL

- Comprehend science materials

GRAPHIC SKILL

- Understand diagrams

GED REVIEW

1. Think About the Topic

The program that you are about to watch is about *Life Science*. Because this area of science is essential to helping us understand the natural world, Life Science questions make up about one-half of the GED Science Test.

This program is about the science of living things, which include the cell (the smallest unit of life), plants, animals (including humans), and the entire natural environment.

To help you understand the importance of life science, this program will take you to a science lab, a farm that preserves traditional breeds of animals, and an artificial coral reef.

2. Prepare to Watch the Video

On the program, you will get an overview of the life sciences through interviews with scientists, researchers, and teachers. Write one activity that you think of when you picture each of the scientists below.

Medical researcher _____

Environmental researcher _____

Paleontologist (studies fossils) _____

You might have said something like: *a medical researcher finds cures for diseases, an environmental researcher helps to save wildlife,* and *a paleontologist researches extinct animals.*

3. Preview the Questions

Look over the questions under *Think About the Program* below and keep them in mind as you watch the program. After you watch, use the questions to review the main ideas in the program.

4. Study the Vocabulary

Review the terms to the right. Understanding the meaning of key life science vocabulary will help you understand the video and the rest of this lesson.

WATCH THE PROGRAM

As you watch the program, pay special attention to the host who introduces or summarizes major life science ideas that you need to learn about. The host will also give you important information about the GED Science Test.

AFTER YOU WATCH

16

1. Think About the Program

What do you think is one reason to learn about cells?

What can be a benefit of learning about human genetics? What can be a danger of genetic research?

What is one way that humans have helped the environment?

What is one way that humans have caused problems with the environment?

2. Make the Connection

The program talks about genetic testing. Would you take a genetic test to see if you had inherited a harmful condition? Why or why not?

TERMS

cell—the basic unit of life; all living things are made of cells

ecology—the study of the relationship of living things to one another and the environment

ecosystem—a community of organisms in their environment

environment—the living and nonliving things that surround an organism

evolution—the gradual changes in species over time

genetics—the study of traits and how they are passed on to offspring

meiosis—a type of cell division in which each resulting cell receives half the genetic material of the original cell

mitosis—a type of cell division in which each resulting cell receives complete genetic material identical to that of the original cell

photosynthesis—the process green plants use to convert light energy, water, and carbon dioxide into food

Cells—The Building Blocks of Life

What Are Cells?

As you learned in the program, all living things are made of **cells,** the basic unit of life. Some living things, like animals and plants, are made of millions of cells. Such organisms have many different types of cells. In humans, for example, there are blood cells, skin cells, and bone cells. Each looks different and performs a different job. In a plant there are different types of cells in the leaves, roots, stems, and flowers. Other living things are made of only one cell. For example, **bacteria** are one-celled organisms.

Most cells are tiny. You cannot see them with your eyes alone. Instead, you need a microscope to see them. However, some cells are quite large. In fact, you probably have a few of these large cells in your refrigerator. Eggs are the reproductive cells of chickens. Each egg is a single, large cell. The largest cell is the ostrich egg. It is about 20 inches across!

Although there are many types of cells, all cells have several things in common:
- Cells use energy.
- Cells move and/or have parts that move.
- Cells can grow, divide, and die.
- Cells have an outer covering, a jellylike inside, and genetic material, which stores information.

The Structure of Cells

All cells have an outer covering called the **cell membrane.** The cell membrane protects the cell from the environment around it. It allows certain things, such as water and food, to enter the cell. It allows other things, such as wastes, to leave the cell. Plant cells also have a **cell wall** outside the cell membrane. The cell wall is tough. It helps give the plant shape and support it. Animal cells do not have cell walls.

Inside most cells is a large structure called the **nucleus.** The nucleus is the control center of the cell. It contains the cell's genetic material, which controls the cell's activities. In some cells, like bacteria, there is no nucleus. Instead, the genetic material occurs as a single strand that forms a ring inside the cell. For more information about the genetic material called DNA, see Science Resources, page 122.

The rest of a cell is mostly **cytoplasm.** This is a thick, watery material like jelly. It surrounds the nucleus and contains the cell's other structures.

Look at the diagram of a typical plant cell. You can see the cell membrane and cell wall, which protects the cell from its surroundings. The cytoplasm contains several other structures:

- Mitochondria (plural of *mitochondrion*) break down food to make energy. They are often called the cell's "powerhouses."
- The chloroplasts help plants make food. Animal cells do not have chloroplasts.
- A vacuole is a storage area. Most plant cells have one large vacuole full of water. Some animal cells do not have vacuoles.

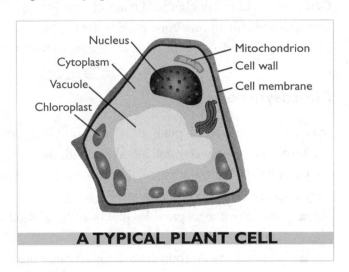

Nucleus
Cytoplasm
Vacuole
Chloroplast
Mitochondrion
Cell wall
Cell membrane

A TYPICAL PLANT CELL

LIFE SCIENCE ▪ PRACTICE I

A. Based on the information on page 20 and the picture above, write *True* or *False* next to each statement.

_____ **1.** The cell is the basic unit of all living things.

_____ **2.** All the cells in the human body are alike.

_____ **3.** Most cells are so tiny they can be seen only with a microscope.

_____ **4.** The nucleus is the control center of a cell.

_____ **5.** The mitochondria are called powerhouses because they produce energy from food.

_____ **6.** The chloroplasts surround the plant cell's cytoplasm.

_____ **7.** A bacterial cell has no nucleus; the genetic material occurs as a simple ring inside the cell.

_____ **8.** All cells are capable of growing and dividing.

B. Describe two ways that plant cells differ from animal cells.

9. _____

_____

10. _____

_____

Answers and explanations start on page 110.

Cell Functions

Cells carry out many chemical reactions to live, grow, and reproduce. Two of the most important chemical reactions in cells are (1) photosynthesis, which takes place in green plants, algae, and some bacteria, and (2) respiration, which takes place in all cells.

Photosynthesis

If you forget to water a plant, or if you put it in a dark closet, the plant will die. That's because plants use water and air, along with the energy from light, to make food energy. As you learned from the video, this process is called **photosynthesis.** Photosynthesis is very important for several reasons:

- Photosynthesis provides plants with the food energy they need to grow and reproduce.
- Plants provide food energy for all other organisms on Earth. Other organisms get energy either by eating plants or by eating organisms that have eaten plants.
- Photosynthesis helps ensure the air has the correct balance of oxygen and carbon dioxide to sustain life.

Photosynthesis takes place inside the **chloroplasts** of plant cells. The chloroplasts give plants their green color, and they are plentiful in plants' leaves.

Photosynthesis takes place in two stages:

1. In the first stage of photosynthesis, the chloroplasts absorb energy from light.

2. In the second stage of photosynthesis, the energy gained from the light is used to make food. The plant uses water absorbed by its roots and carbon dioxide absorbed from the air as raw materials. The food is produced in the form of sugars. Most of the oxygen that is also produced passes into the air.

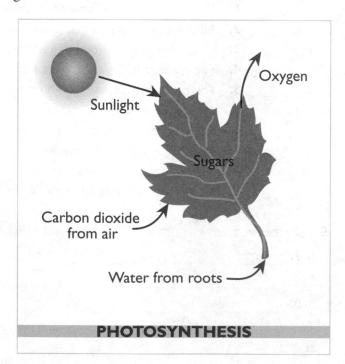

PHOTOSYNTHESIS

Respiration

Unlike photosynthesis, respiration takes place in the mitochondria of all cells. **Respiration** is the process by which cells release energy from sugars. The waste products—water vapor and carbon dioxide—are released into the air. Because this process takes place inside the cell, it is often referred to as cellular respiration.

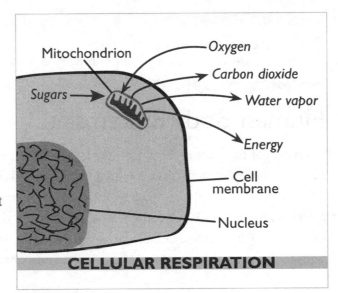

CELLULAR RESPIRATION

LIFE SCIENCE ▪ PRACTICE 2

A. Use the information and diagrams on pages 22 and 23 to fill in the chart below.

Chemical Reaction	Where It Takes Place	Raw Materials Needed	What Is Produced
Photosynthesis	In the cells of green plants, algae, and some bacteria	Energy from light 1. _____ 2. _____	3. _____ _____ Oxygen
Respiration	4. _____	5. _____ 6. _____	Energy 7. _____ 8. _____

B. Answer the following question based on the information on pages 22 and 23.

9. What would happen if you placed a green plant in a dark place? Why?

_____

_____

_____

Answers and explanations start on page 110.

"In your body you have many different kinds of cells equipped to carry out specific functions—blood cells, brain cells, skin cells, to name just a few."

Human Body Systems

Each cell of the human body has a task to perform, and together the cells carry out all the complicated functions of a human being, from breathing and walking to digesting and thinking. The cells of the human body are organized on several levels. For example, a group of nerve **cells,** or neurons, makes up nerve **tissue.** Nerve tissue and other types of tissue, such as blood vessels, make up the brain, an **organ.** And the brain, together with the spine and nerves, make up the **nervous system.**

The Nervous System

The **nervous system** is the information-processing and control system of the body. It gathers information about activities inside the body, and it also gathers information about the outside environment. The nervous system then processes the information and directs the body to respond. For example, if the temperature drops, nerves on your skin send this information to the brain. The brain then directs you to shiver, which warms you. It also causes you to decide to put on warmer clothing.

Parts of the Nervous System

As you can see in the diagram, the nervous system has two main parts: the central nervous system and the peripheral nervous system.

- The **central nervous system** consists of the brain and the spinal cord. It is the control center of your body.
- The **peripheral nervous system** consists of a network of nerves. These nerves connect all parts of the body to the central nervous system.

The nervous systems contain billions of nerve cells, called **neurons.** Neurons carry messages rapidly from the peripheral nervous system to the central nervous system and back.

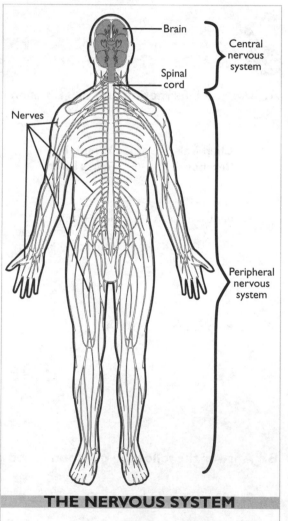

THE NERVOUS SYSTEM

The Brain

The brain is the control organ of the central nervous system. It has three main parts:

- The **cerebrum** controls functions such as speaking, thinking, seeing, and hearing.
- The **cerebellum** controls balance and coordination.
- The **brain stem** controls heartbeat, breathing, and other life functions.

The brain is protected by three outer membranes, called the meninges, and the skull. You may have heard of the disease meningitis. It is a serious infection of the meninges.

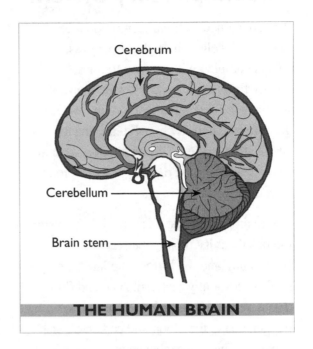

Cerebrum

Cerebellum

Brain stem

THE HUMAN BRAIN

LIFE SCIENCE ■ PRACTICE 3

A. Use the information and diagrams on pages 24 and 25 to answer the following questions. Place the letter of the correct answer in the space provided.

_____ 1. Which part of the nervous system connects parts of the body to the brain and spinal cord?
- **a.** the central nervous system
- **b.** the peripheral nervous system
- **c.** the cerebellum
- **d.** the cerebrum

_____ 2. Which part of the nervous system is the control center for the body?
- **a.** the neurons
- **b.** the meninges
- **c.** the central nervous system
- **d.** the peripheral nervous system

_____ 3. One function of the brain stem is to control _____
- **a.** thinking
- **b.** speaking
- **c.** breathing
- **d.** balance

B. Use the information on pages 24 and 25 to answer the following question.

4. In what order should the following terms be arranged if you put the simplest unit first and the most complex unit last? Place the numbers 1 through 4 next to each term in the proper order.

_____ cell   _____ system   _____ tissue   _____ organ

Answers and explanations start on page 110.

The Respiratory System

As you learned, cellular respiration takes place in all cells. Through this process, sugars combine with oxygen to release energy for the cell's use. How do humans get oxygen for cellular respiration? We get it from the air by breathing, using the **respiratory system.** The respiratory system also releases carbon dioxide, a waste product of cellular respiration, into the air.

The main parts of the respiratory system are the **nasal cavity** (nose), the **trachea** (windpipe), and the **lungs.** The lungs are made of a spongy material. As blood flows through the lungs in tiny blood vessels, oxygen enters the blood and carbon dioxide leaves it. This gas exchange takes place in tiny air sacs called **alveoli.**

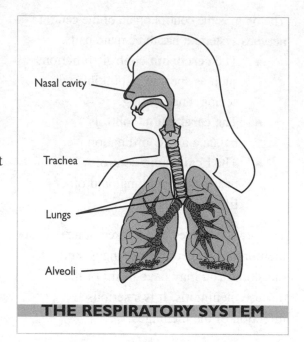

Nasal cavity

Trachea

Lungs

Alveoli

THE RESPIRATORY SYSTEM

The Circulatory System

The **circulatory system** carries oxygen, nutrients, and other substances to cells throughout the body. It also removes wastes such as carbon dioxide from the cells. The circulatory system consists of the **heart,** the **blood vessels,** and the **blood** itself. It is sometimes called the cardiovascular system.

The main organ of the circulatory system is the heart. It is a pump made of muscle tissue. The average heart beats 70 to 80 times a minute and pumps 5 quarts of blood a minute. During hard exercise, the heart may pump up to five times as much blood per minute!

There are several types of blood vessels. **Arteries** are large blood vessels that carry blood away from the heart. These branch into smaller and smaller vessels. The tiniest blood vessels are the **capillaries.** In the capillaries, substances can move into and out of the blood through the thin capillary walls. From the capillaries, blood flows into small **veins,** larger veins, and back to the heart.

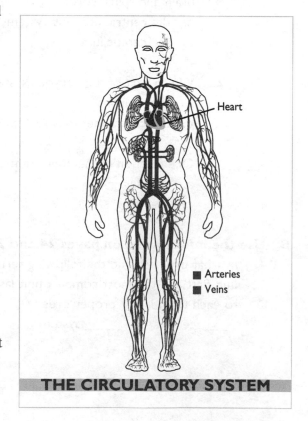

Heart

■ Arteries
■ Veins

THE CIRCULATORY SYSTEM

Blood is made of a colorless liquid called **plasma.** Plasma is made mostly of water, but it also contains sugars, salts, proteins, and other substances. In the plasma are three types of blood cells:

- red blood cells, which carry oxygen around the body
- white blood cells, which help defend the body against disease
- platelets, which play a role in clotting when a blood vessel is damaged

The average adult has about 10.5 pints of blood. After you give a pint of blood, you must drink a lot of water so your body can replenish your blood supply.

LIFE SCIENCE ■ PRACTICE 4

A. Refer to the information and diagrams on pages 26 and 27 to fill in the word or phrase that will correctly complete each statement.

1. The main function of the respiratory system is to absorb _____ from the air into the blood, and to remove _____ from the blood and release it into the air.

2. When you breathe in, air flows into the nose, also called the _____, through the mouth, the throat, and down the _____, also called the windpipe.

3. In the lungs, the gas exchange between the air and the blood takes place in the _____.

4. The main function of the circulatory system is to carry _____, _____, and other substances to the body's cells and to remove _____ and other wastes.

5. The main organ of the circulatory system is the _____, which pumps blood throughout the body.

6. _____ carry blood away from the heart, and _____ carry blood to the heart.

7. Blood is made mostly of _____ and contains three types of blood cells.

B. Use the information on pages 26 and 27 to answer the following question.

8. How are the functions of the respiratory and circulatory systems related?

_____

Answers and explanations start on page 111.

PROGRAM 16 ■ LIFE SCIENCE

"Every organism...depends on every other one; otherwise they're not going to stay alive."

Ecosystems and the Environment

What Is an Ecosystem?

Every living thing on Earth, including humans, lives in an ecosystem. An **ecosystem** is a community of plants, animals, and other organisms. It includes the physical **environment** in which they live—the air, soil, water, and climate. The organisms in an ecosystem interact with each other and with the environment.

Some ecosystems are very large, such as the ocean. Others are quite small, such as a freshwater pond or a meadow. In each ecosystem, there is a variety of life forms. There are bacteria, plants, fungi, insects, mammals, birds, and so on. The variety of life is called **biodiversity.**

As Professor Smith explained on the video, the organisms in an ecosystem depend on one another and on the natural environment to survive. There are many types of relationships among organisms in an ecosystem. Here are a few:

- **Organisms compete for food, water, and shelter in an ecosystem.** For example, two types of birds may compete for the best nesting areas in the woods.
- **Some organisms cooperate with one another, and both benefit.** For example, lichens consist of a fungus and an alga living together. Through photosynthesis, the alga makes food for itself and the fungus. The fungus, which cannot make its own food, provides water for the alga.
- **Some organisms have a predator-prey relationship.** That means that the **predator** kills and eats the **prey.** For example, a fox may kill and eat a squirrel.

The Flow of Energy in an Ecosystem

As you recall, plants get energy by changing the energy from sunlight into food energy through the process of photosynthesis. Then other organisms get their energy by eating plants or by eating organisms that eat plants. Thus, energy flows through an ecosystem, from the sun to plants to other organisms. The organisms in an ecosystem can be divided into three basic types according to the role they play in energy flow:

- **Producers** are organisms that carry out photosynthesis to get food energy. Green plants, algae, and some bacteria are producers.
- **Consumers** are organisms that eat other organisms to get food energy. All animals are consumers.
- **Decomposers** are organisms that get food energy from eating dead organisms. Most fungi and many bacteria are decomposers.

One way to show the flow of energy in an ecosystem is to use a diagram called a **trophic pyramid.** *Trophic* means "feeding," so a trophic pyramid shows feeding relationships. The producers of an ecosystem, such as plants, are always at the bottom level. At the next level are primary consumers, organisms that feed on the producers. At the third level are secondary consumers, organisms that feed on the primary consumers. There may be fourth and fifth levels as well.

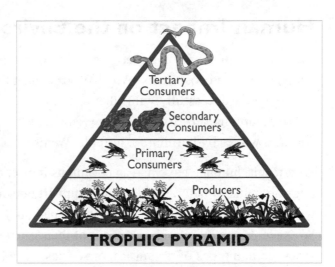

TROPHIC PYRAMID

The diagram above is a simple trophic pyramid. Note the wide base occupied by the producers. As the trophic level gets higher, the pyramid gets smaller. This shows that the amount of food energy available at each level is decreasing. It is decreasing because organisms are consuming it and because it is being lost to the air as heat energy. Thus, the number of organisms at each higher level decreases, too—there is less food energy for them. As you can see, in this trophic pyramid, there are many plants but few snakes.

LIFE SCIENCE • PRACTICE 5

A. Refer to the information on page 28, and write the letter of the correct term next to its definition.

_____ **1.** an organism that gets food energy from other organisms

_____ **2.** the air, soil, water, and climate of an area

_____ **3.** a community of organisms and its physical surroundings

_____ **4.** an organism that performs photosynthesis to make food energy

_____ **5.** an organism that hunts, kills, and eats another organism

_____ **6.** the variety of types of organisms in an area

a. predator
b. environment
c. producer
d. ecosystem
e. biodiversity
f. consumer

B. Refer to the trophic pyramid above to answer the following questions.

7. What might happen if a disease killed most of the frogs in this ecosystem?

_____

8. What might happen if a fire wiped out all the plants in the ecosystem?

_____

Answers and explanations start on page 111.

Human Impact on the Environment

Humans live in almost every ecosystem on Earth. From the Arctic to the Equator, humans rule the ecosystems. Why? We can adapt to a wide variety of environments. We compete successfully with other species for the resources we need. We have learned to use nature, not just live in it. For example, we grow food rather than hunt it. We keep warm in winter by burning fossil fuels. We use tools and technology.

As a result, humans have had an enormous impact on the natural environment. We use huge amounts of natural resources to support ourselves. Because there are so many of us, we have spread all over the world. In our path, other organisms have not done as well. We have driven many organisms out of their usual homes—their **habitats.** In many cases, certain types of organisms, or species, have been wiped out entirely.

Loss and Destruction of Habitat

What human activities cause habitat loss and destruction? Here are a few:

- *Farming.* Almost 45 percent of all U.S. land is used to raise crops or graze animals. Farmers' chemicals and animal wastes sometimes pollute water habitats.
- *Urbanization.* Cities and suburbs have spread over huge areas, especially in California and on the East Coast. They also are a major source of pollution.
- *Logging.* Logging can damage forests and reduce habitat area for wildlife.

Another human activity causes habitat destruction: the introduction of a nonnative species into an ecosystem. If the species has no natural enemies in the ecosystem, it can take over, driving out the original organisms. For example, the opossum shrimp was introduced to lakes in Montana. People thought these shrimp would be food for native salmon. After all, the salmon ate native species of crustaceans in the lakes. People reasoned that more shrimp for salmon would result in more salmon to catch and sell. However, the shrimp ate the native crustaceans, and the salmon's main source of food was destroyed. The decrease in the number of salmon led to a decrease in the number of animals that eat salmon, such as bald eagles and grizzly bears. The ecosystem's balance was so upset it was no longer a good habitat for many of its native species.

Effects of Habitat Loss

Loss of habitat is the leading cause of population decline among many species. In 1973, Congress passed the **Endangered Species Act.** Its purpose is to protect species that are at risk from human activity. According to the U.S. Fish and Wildlife Services, a species is "endangered" if it faces **extinction,** or being wiped out, in most or all of its habitat. It is "threatened" if it is likely to become endangered in the near future. In some cases, there are plans to rescue species, such as the condor, a large bird. The chart on the next page shows the number of species officially listed under the Endangered Species Act.

Group	Endangered Species	Threatened Species	Recovery Plans
Mammals	65	9	53
Birds	78	14	75
Reptiles*	14	22	32
Amphibians	11	8	12
Fish	71	44	95
Invertebrate animals (clams, snails, insects, arachnids, crustaceans)	148	31	129
Flowering Plants	568	144	556
Conifers and cycads*	2	1	2
Ferns	24	2	26
Lichens*	2	0	2
Total	983	275	982

*not shown

ENDANGERED OR THREATENED SPECIES IN THE UNITED STATES, 2002

Source: U.S. Fish and Wildlife Service

LIFE SCIENCE ▪ PRACTICE 6

A. Refer to the information and the chart on pages 30 and 31 to fill in a word or phrase that correctly completes each statement.

1. The area in which a particular species can live is its _____.

2. Three human activities that harm the environment are _____, _____, and _____.

3. If a nonnative species has no _____ in its new ecosystem, it may take over.

4. The Endangered Species Act tries to prevent species from becoming _____.

5. According to the chart, _____ bird species were listed as endangered in 2002.

6. According to the chart, 144 species of _____ were listed as threatened in 2002.

B. Use the information on pages 30 and 31 to answer the following question.

7. People who want to build homes or industry sometimes come into conflict with people who want to save wildlife. Why is this so?

    _____

Answers and explanations start on page 111.

Comprehend Science Materials

When you comprehend what someone is saying to you, you understand what the person tells you. You understand their main point and any supporting facts or details. On the GED Science Test, some questions will test your comprehension of scientific facts or concepts.

When you **comprehend** something, you understand what is stated or shown in a picture. You may restate a fact in other words, summarize a passage or a graphic, or identify the implications of what is stated. You also may need to **infer**—or figure out something that is suggested but not directly stated in the passage.

EXAMPLE

Lab-grown skin cells have been used to help people with severe skin problems. Now scientists are growing other types of cells. For example, they are growing cartilage, the flexible material that supports the nose and ears. They can even grow custom-made cartilage. The process starts with a scaffold that looks like attached spaghetti strands. The scaffold is cut and molded to the right shape— let's say, the patient's left ear. Then young cartilage cells are placed on the scaffold. The scaffold is then placed under the skin of a lab mouse. The mouse provides food and oxygen for the cells as they grow. Eventually, a new ear forms.

1. What is cartilage? _____

_____

Did you say something like: *Cartilage is a flexible material in the body. Ears and noses are supported by cartilage?* If so, you are correct.

THINKING STRATEGY: Quickly skim the passage to find the word or phrase. Then reread the entire sentence or sentences with the word or phrase to figure out the answer.

2. How might lab-grown cartilage help people in the future? _____

_____

If you said that *in the future, lab-grown cartilage may replace damaged cartilage in patients,* you are right. You inferred something from what is directly stated in the passage. Someday this process may be more than a lab experiment.

Now let's look at comprehension questions similar to those on the GED Science Test.

Sample GED Question

Seven out of ten adults don't get regular exercise, and four out of ten never exercise at all. These figures were reported by the National Center for Health Statistics, which surveyed 68,000 Americans. Officials find these numbers alarming. People who don't exercise are more likely to be overweight. And people who are overweight are more likely to have serious conditions like diabetes, heart disease, and strokes. Thus, many Americans may be risking their health by not exercising regularly.

Which statement best summarizes this information?

(1) Many Americans' health is at risk because of lack of exercise.

(2) Most American adults get little or no exercise.

(3) The National Center for Health Statistics surveyed 68,000 Americans.

(4) Many Americans have diabetes, heart disease, and strokes.

(5) Health officials find exercise statistics alarming.

THINKING STRATEGY: A summary is a brief statement of the main points. Often the topic sentence of a paragraph gives a good summary of the paragraph's main idea. Look for the topic sentence of the paragraph.

The correct answer is **(1) Many Americans' health is at risk because of lack of exercise.** This answer summarizes the paragraph's main point by restating the last sentence, which is the topic sentence of the paragraph.

GED THINKING SKILL PRACTICE ▪ COMPREHENSION

Questions 1 and 2 are based on the passage below.

In Japan, monkeys have been moving from wilderness areas to farm areas and towns, where food is plentiful. Monkeys eat crops and garbage. In fifty years, the monkey population grew from 15,000 to 150,000.

1. Based on the passage, you can infer that the monkeys are moving

   (1) to zoos

   (2) back to the wilderness

   (3) to picnic areas

   (4) near people

   (5) to national parks

2. The passage suggests the monkey population grew because

   (1) there was no more wilderness

   (2) they were treated like pets

   (3) food was easy to get

   (4) farmers raised monkeys

   (5) more baby monkeys were born

Answers and explanations start on page 112.

Understand Diagrams

On the GED Science Test, you will answer questions based on **diagrams** similar to the ones that you saw on the video. Some diagrams show processes—for example, how energy, in the form of food, passes through a community of plants and animals, as in a simple food web. Others show how things look, such as the parts of an insect.

EXAMPLE

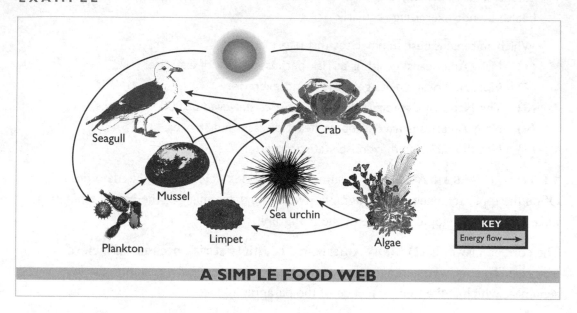

A SIMPLE FOOD WEB

1. How do limpets get their energy in this food web?

_____

Did you say *By eating algae*? That's right. The **title** indicates you are looking at food relationships. The **key** tells you that the arrows show the flow of energy. The arrow pointing to the limpet comes from the algae. Thus, limpets get energy from algae.

When you are asked a question about a diagram, first read the title and examine the key, if there is one. Read the labels that identify parts. Finally, locate the detail you need.

2. The title of this diagram is "A Simple Food Web." What does this title suggest about the energy relationships in a larger ecosystem, for example, a seacoast? Give a reason to support your answer.

_____

_____

Your answer should have been something like: *The energy relationships on the seacoast would be more complicated, since many more types of organisms live there and feed on one another.* The word *simple* in the title means the food web shown is not complex.

Now let's look at diagrams similar to those you will see on the GED Science Test.

GED Diagram Practice

The word *heartburn* describes the burning feeling near the heart that you get after you eat something that disagrees with you.

According to the diagram, what is the actual source of the burning feeling in heartburn?

(1) fatigue in the heart muscles

(2) buildup of acid in the heart muscles

(3) acid flowing into the stomach

(4) acid rising into the esophagus from the stomach

(5) acid remaining in the stomach

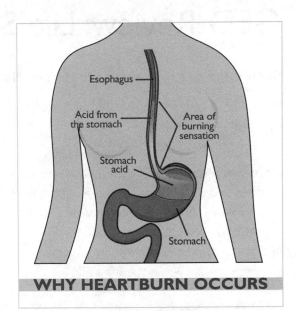

WHY HEARTBURN OCCURS

THINKING STRATEGY: First read the short passage. Then examine the details in the diagram to find what might be causing the burning sensation of heartburn. Look at the title, the labels, and the arrow.

The correct answer is **(4) acid rising into the esophagus from the stomach.** The passage explains that heartburn can result from eating food that is hard to digest. The arrows in the diagram show that when acid backs up out of the stomach, it rises up into the esophagus, or food pipe. This is the irritated area with the burning feeling.

GED GRAPHIC SKILL PRACTICE ▪ USING DIAGRAMS

Questions 1 and 2 are based on the information and diagram below.

The male sex cell is called a sperm cell. Millions of sperm cells try to reach a female sex cell, or egg. They whip their tails back and forth in order to move. The sperm cell that wins this race fertilizes the egg.

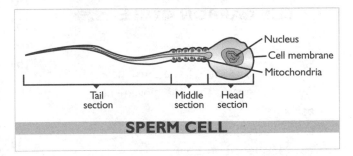

SPERM CELL

1. Into how many sections is a sperm cell divided?

   (1) two

   (2) three

   (3) four

   (4) five

   (5) six

2. A sperm cell moves toward an egg cell by

   (1) splitting its nucleus

   (2) spinning its head

   (3) expelling its mitochondria

   (4) allowing water into its head

   (5) flicking its tail back and forth

Answers and explanations start on page 112.

GED Review: Life Science

Choose the <u>one best answer</u> to the questions below.

<u>Questions 1 and 2</u> refer to the following information and diagram.

In the carbon cycle, the element carbon moves from one part of the environment to another and back again. In the atmosphere, it is found in the gas carbon dioxide (CO_2). Plants and animals move carbon into and out of the atmosphere through several processes as shown in the diagram.

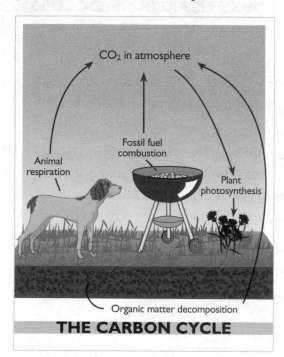

CO_2 in atmosphere

Fossil fuel combustion

Animal respiration

Plant photosynthesis

Organic matter decomposition

THE CARBON CYCLE

1. The diagram's title, "The Carbon Cycle," suggests that the movement of carbon
   (1) occurred only when life first developed on Earth
   (2) is an ongoing process that never stops
   (3) is the only natural cycle on Earth
   (4) is vital to the continuance of life on Earth
   (5) will stop once all the carbon reaches the atmosphere

2. According to the diagram, carbon is removed from the atmosphere by
   (1) organic matter
   (2) respiration
   (3) decomposition
   (4) combustion
   (5) photosynthesis

<u>Questions 3 and 4</u> refer to this information.

Scientists have altered the genes of mice to produce extra CaMK. CaMK is a protein that helps muscle cells improve their strength. The protein works by increasing the number of mitochondria in each cell. As a result of the extra CaMK, the muscle cells of the altered mice became stronger without exercise. The altered, inactive mice became as strong as mice that were not altered but that were made to exercise.

3. What is CaMK?
   (1) a mitochondrion
   (2) a type of mouse
   (3) a protein
   (4) a gene
   (5) a muscle

4. What resulted from genetically altering the mice to produce extra CaMK?
   (1) The mice became stronger without exercise.
   (2) The mice became stronger with exercise.
   (3) The mice were weaker than mice that exercised.
   (4) The mice produced fewer muscle cells.
   (5) The mice were damaged.

The outer ear collects sound and sends it down the auditory canal to the eardrum. The eardrum then vibrates tiny bones called the hammer, anvil, and stirrup. In turn, this moves the fluids in the cochlea. Finally, this activates the auditory nerve, which sends signals to the brain. The brain interprets these signals as sound.

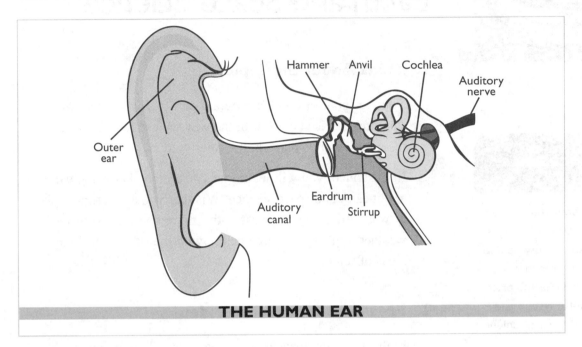

THE HUMAN EAR

5. Which part of the ear collects sounds?
   (1) the auditory nerve
   (2) the cochlea
   (3) the hammer
   (4) the auditory canal
   (5) the outer ear

6. Which part of the ear does a doctor check by looking inside with an instrument?
   (1) the anvil
   (2) the eardrum
   (3) the cochlea
   (4) the auditory nerve
   (5) the stirrup

Question 7 refers to the following information.

Coral reefs cover less than 1 percent of the ocean's area. Yet they are home to 33 percent of all fish species. Coral reefs provide fish for eating, protect coastal areas, and attract tourists. As valuable as coral reefs are, a 1998 study estimated that 58 percent of the world's coral reefs are threatened by human activity.

7. What percentage of coral reefs are threatened by human activity?
   (1) less than 1 percent
   (2) about 33 percent
   (3) close to 60 percent
   (4) 98 percent
   (5) almost 100 percent

Answers and explanations start on page 112.

Earth and Space Science

LESSON GOALS

SCIENCE SKILLS

- Learn about the different ways that Earth changes
- Think about how people use Earth's resources
- Understand Earth's place in the solar system and what makes Earth unique

THINKING SKILL

- Apply science information

GRAPHIC SKILL

- Understand graphs

GED REVIEW

1. Think About the Topic

You are about to watch a program on *Earth and Space Science.* The program is about Earth, the way we use its natural resources, and its place in the solar system.

In this video, you will learn that you already know a great deal about earth and space science. Whether you are getting a glass of water, filling up your car with gas, or checking your local weather forecast, you are dealing with the subject of these areas of science.

2. Prepare to Watch the Video

In this program, teachers and scientists will give you an overview of earth and space science. You will learn how scientists use their knowledge of Earth and how it works, not only to learn about the past, but also to make predictions about the future. Why might it be important to be able to make future predictions about things like earthquakes and storms?

_____

_____

_____

_____

You may have written something like: *If we can predict when and where earthquakes, tornadoes, or hurricanes will occur, we can warn people earlier. We can also prepare properly to save lives and lessen the damage done to communities.*

3. Preview the Questions

Look over the questions under *Think About the Program* below and keep them in mind as you watch the program. After you watch, use the questions to review the main ideas in the program.

4. Study the Vocabulary

Review the terms to the right. Understanding the meaning of key earth and space science vocabulary will help you understand the video and the rest of this lesson.

WATCH THE PROGRAM

As you watch the program, pay special attention to the host who introduces or summarizes major earth and space science ideas that you need to learn about. The host will also give you important information about the GED Science Test.

AFTER YOU WATCH

1. Think About the Program

Why is it important to collect and keep data on earthquakes?

What are some of the reasons that Earth is capable of supporting life?

What can scientists learn from studying the rock cycle?

Can the weather be predicted with complete accuracy?

2. Make the Connection

The program discusses nonrenewable resources. Scientists are working to conserve some of our nonrenewable resources by developing new technologies, such as electric- and solar-powered cars. Would you choose to buy one of these types of cars? Why or why not?

air mass—a large body of air uniform in temperature and humidity

climate—average weather conditions over a long period of years

crust—the top layer of Earth

earthquake—a slippage in underground rocks that causes shaking on the surface

fault—a break between two or more blocks of rock along which the rock blocks move

fossil fuel—fuel that formed from remains of plants and animals; fossil fuels include petroleum, natural gas, and coal

nonrenewable resource—a resource that can be used up

plate tectonics—a theory that explains how Earth's crust changes

precipitation—rain, sleet, and snow

renewable resource—a resource in endless supply

solar system—the sun and the objects that circle it

volcano—a place where molten rock comes out onto Earth's surface

water cycle—the movement of water from the atmosphere to the surface and back

"Most Earth changes are gradual.... Others have the ability to alter the face of the Earth and people's lives in mere seconds."

The Changing Earth

Earthquakes and Volcanoes

During an **earthquake,** you can feel the ground swaying and shaking. If the earthquake is severe, it can change the landscape and bring down buildings, bridges, and highways. If it occurs where people live, many may be injured or killed.

What is an earthquake? It is a sudden slipping of rock along a fault. A **fault** is a break in rock along which the resulting rock blocks move. When pressure builds up underground, it may cause the blocks to move suddenly along the fault. The energy released by the movement causes shaking that is felt on the surface. The energy of an earthquake travels in **seismic waves** right through Earth. These waves can be detected all over the world.

Areas that have frequent earthquakes often have volcanoes as well, as shown in the map below. A **volcano** forms when melted rock, called **magma,** comes up to the surface of Earth from deep underground.

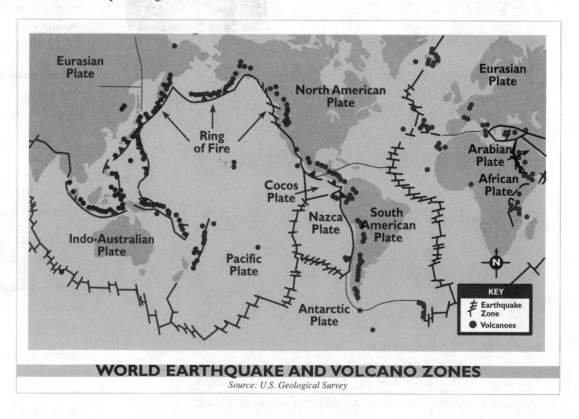

WORLD EARTHQUAKE AND VOLCANO ZONES

Source: U.S. Geological Survey

The Theory of Plate Tectonics

Why do earthquakes and volcanoes occur? The theory of **plate tectonics** tries to answer this question. According to this theory, the **crust,** or outer layer of Earth, is made of about twenty **plates.** These are like jigsaw puzzle pieces, fitting roughly together. The plates move past one another, collide with one another, and move away from one another. As you can see on the map, plate boundaries are the areas where earthquakes and volcanoes are likely to occur. The circle of active volcanoes and earthquake zones around the Pacific Ocean is known as the *Ring of Fire*.

EARTH AND SPACE SCIENCE ▪ PRACTICE 1

A. Use the information and map on pages 40 and 41 to complete these sentences.

1. The energy from an earthquake travels through the ground in _____ waves that can be detected on the other side of Earth.

2. The sudden slipping of large blocks of rock along a _____ causes an earthquake.

3. According to the map, in North America, most volcanoes and earthquakes occur on the _____ coast of the continent.

4. A _____ forms where hot magma breaks through the surface of Earth.

5. According to the theory of plate tectonics, the _____ of Earth is made of about twenty _____ that fit together like a rough jigsaw puzzle.

6. Most earthquakes and volcanoes occur along plate _____.

B. Use the information and map on pages 40 and 41 to answer the following questions.

7. Why is *Ring of Fire* a good name for the earthquake and volcano zone that surrounds the Pacific Ocean?

   _____

   _____

8. Where are most of the plate boundaries located?

   _____

Answers and explanations start on page 112.

Meteorology, the Study of Weather

Changes in the weather result from the changing conditions in the atmosphere—the air that surrounds Earth. Scientists who study weather are called **meteorologists.** As meteorologist Kenny Priddy explains on the video, meteorologists observe and collect a lot of data. They record air temperature, air density, air pressure, winds, clouds, water temperature, and **precipitation** (rain or snow). Based on these data, they predict what the weather will be like tomorrow and the day after. Because weather is so complex, forecasts for more than a few days ahead are generally not very accurate.

Air Masses and Weather

Most weather predictions are based on the movements and characteristics of air masses. An **air mass** is a large area of air that is uniform in its temperature and **humidity** (amount of water vapor). An air mass forms when a large body of air rests over the surface long enough to pick up the conditions of that surface. For example, an air mass that forms over a desert will be dry. An air mass that forms over a body of water will be moist. There are four types of air masses:

- **Continental tropical air masses** form over land near the equator, where it is warm. They are warm and dry.
- **Maritime tropical air masses** form over oceans near the equator. They are warm and humid.
- **Continental polar air masses** form over cold land near the poles. They are cold and dry.
- **Maritime polar air masses** form over cold oceans. They are cold and humid.

As air masses move slowly across Earth's surface, they affect the weather of the areas they are passing over. They also change, becoming warmer or cooler, wetter or drier. Where two air masses meet, a front is formed. The weather at fronts is very changeable. Temperatures may rise or fall quickly. There is often precipitation.

Climate

Climate refers to the average weather conditions in a particular place over a period of many years. There are many ways to classify, or group, climates. One way is by average precipitation and temperature. For example, a tropical desert climate is one in which little precipitation falls and temperatures are very warm. A temperate climate is one with moderate precipitation and temperatures. The graph on the next page shows the average monthly temperature in Louisville, Kentucky.

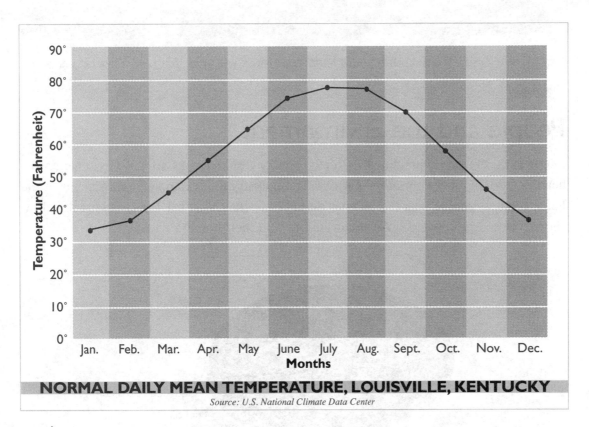

NORMAL DAILY MEAN TEMPERATURE, LOUISVILLE, KENTUCKY

Source: U.S. National Climate Data Center

A. Based on the information on page 42, write *True* or *False* next to each statement. Correct each of the false statements.

_____ 1. Meteorologists can accurately forecast the weather a month ahead.

_____ 2. Both rain and snow are forms of density.

_____ 3. A continental polar air mass forms over cold land.

_____ 4. A maritime tropical air mass is warm and dry.

_____ 5. Climate refers to the average weather conditions in a place over a long period of time.

B. Use the graph above to answer the following questions.

6. Normally, what is the warmest month of the year in Louisville? _____

7. Does Louisville have a tropical or temperate climate? Explain.

_____

_____

_____

Answers and explanations start on page 113.

SCIENCE SKILLS

"Energy is one of the most important resources we take from the Earth system."

People and the Environment

In our daily life, we depend heavily on the energy resources of Earth. We use energy for heating, cooking, transportation, farming, and industry. For example, in most areas of the United States, people heat their homes during the winter months. The circle graph below shows the energy resources Americans use to heat their homes.

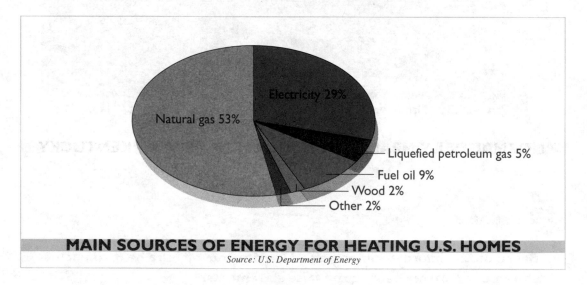

Natural gas 53%

Electricity 29%

Liquefied petroleum gas 5%

Fuel oil 9%

Wood 2%

Other 2%

MAIN SOURCES OF ENERGY FOR HEATING U.S. HOMES

Source: U.S. Department of Energy

Nonrenewable and Renewable Sources of Energy

Energy resources include **fossil fuels,** such as oil, natural gas, and coal. These sources of energy are **nonrenewable**—when we use them up, there will be no more. As you can see in the graph, most U.S. homes use nonrenewable sources of energy for heating. Even people who heat with electricity are probably using nonrenewable resources. That's because most electricity is made by burning fossil fuels.

Energy resources also include wind power, water power, solar (sun) power, and geothermal energy. These sources of energy are **renewable.** That means there is an endless supply of them. Of the energy sources for heating shown on the graph, only wood is a renewable resource. So you can see that very few U.S. homes use renewable energy sources for heating. The "Other" category on the graph includes the few homes that use solar energy by collecting it in solar cells on roofs.

The main use of renewable energy sources is to generate electricity. About 10 percent of U.S. electricity comes from water, wind, solar power, and other renewable resources.

Energy Resources for the Future

Our use of energy grows each year. Scientists disagree about how long our supply of nonrenewable energy resources will last. Estimates range from a few hundred to a few thousand years. Some people favor more drilling for fossil fuels to increase the supply. Others favor decreasing our use of fossil fuels through conservation and more efficient machines. For example, technologies like the hybrid car and fuel cells can decrease our use of fossil fuels.

EARTH AND SPACE SCIENCE ▪ PRACTICE 3

A. Use the information on page 44 to complete the following chart.

Nonrenewable Sources of Energy	Renewable Sources of Energy
Oil	3. _____
1. _____	4. _____
2. _____	5. _____
	6. _____

B. Use the information and graph on pages 44 and 45 to answer the following questions.

7. According to the graph, what is the most popular source of energy for heating homes in the United States?

_____

8. Of the energy sources shown on the graph, which one is renewable? Why?

_____

9. What are most renewable sources of energy used for in the United States?

_____

10. Why is the use of renewable sources of energy likely to increase in the future?

_____

Answers and explanations start on page 113.

Air Pollution

One result of burning fossil fuels is that air pollution has increased. With increased air pollution has come an increase in breathing problems such as asthma, chronic bronchitis, and other respiratory diseases.

In fact, air pollution became so bad in many areas that Congress passed the Clean Air Act in 1990. As a result of the act, the federal government set new standards for air pollutants such as ozone, carbon monoxide, and particulate matter. Power plants and factories had to decrease the amount of pollution they released into the air. Cities and states began to keep track of the amount of pollutants in the air.

The Air Quality Index

Today we have a system to report and forecast air pollution. It's called the **air quality index.** You may have heard a TV or radio weather report that included the air quality index. For example, the meteorologist may have reported a Code Orange alert for ozone. As you can see from the chart below, that means many people should limit the time they spend outdoors. They should limit the activities they perform outdoors.

INDEX	AQI Nos.	Air Quality	Recommended Actions
Code Green	0 to 50	Good	None
Code Yellow	51 to 100	Moderate	Unusually sensitive people should consider limiting outdoor exertion.
Code Orange	101 to 200	Unhealthy for sensitive groups	Active children and adults and people who have trouble breathing or have respiratory disease such as asthma should limit prolonged or heavy outdoor exertion.
Code Red	201 to 300	Unhealthy	Everyone, especially children, should limit heavy or prolonged outdoor exertion.

AIR QUALITY INDEX

Source: U.S. Environmental Protection Agency

The air quality index also includes Code Purple and Code Maroon, but these high pollution levels are rare. Because people have been working to clean up the air, it's been years since we have had a Code Maroon!

Ground-Level Ozone

You may have heard that the **ozone layer** high in the atmosphere protects us from the harmful ultraviolet rays of the sun. That's true. However, when ozone is at ground level, it is a pollutant. People with respiratory diseases have trouble breathing when there is a Code Orange or Code Red alert for ozone.

In addition, children are more at risk for developing ozone-related health problems, such as asthma or chronic colds. That's because they often play outside during the summer, when ozone levels are the highest. They breathe faster than adults, and they inhale more pollutants relative to their body weight than adults do. On Code Orange and Code Red ozone alert days, children should play indoors.

EARTH AND SPACE SCIENCE ▪ PRACTICE 4

A. Use the information and chart on pages 46 and 47 to answer the following questions. Place the letter of the correct answer in the space provided.

_____ **1.** What was one result of the Clean Air Act of 1990?
   a. new air pollutant standards   **b.** increased pollution from industry
   c. destruction of the ozone layer   **d.** increased pollution in homes

_____ **2.** What is used to report on and forecast air pollution levels?
   a. the Clean Air Act   **b.** weather reports
   c. the air quality index   **d.** pollen count

_____ **3.** Yesterday the air quality index was 75. What was the air quality index rating yesterday?
   a. Code Green   **b.** Code Yellow
   c. Code Orange   **d.** Code Red

_____ **4.** You hear on TV that there is a "bad air" alert. Everyone is urged to stay indoors if possible and limit outdoor activity. What is the air quality index rating likely to be today?
   a. Code Green   **b.** Code Yellow
   c. Code Orange   **d.** Code Red

_____ **5.** Ground-level ozone amounts tend to be the highest during the
   a. winter   **b.** spring
   c. summer   **d.** autumn

B. Use the information and chart on pages 46 and 47 to answer this question.

6. Why does the air quality index warn against heavy exertion (activities that require lots of energy) on days when the air quality is unhealthy? (_Hint:_ Think about what happens to your breathing when you do heavy exercise.)

_____

_____

Answers and explanations start on page 113.

"What makes Earth unique in the solar system? It's the only planet known to support life."

The Solar System and the Universe

The **solar system** consists of the sun, the nine major planets and their satellites, minor planets (also called asteroids), comets, and dust and debris. The sun has more than 99 percent of the mass (matter) in the solar system. Thus, it is at the center of the solar system, and its gravitational pull holds everything else in orbit around it.

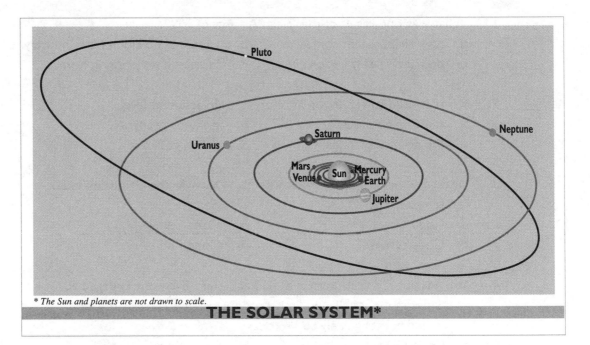

* The Sun and planets are not drawn to scale.

THE SOLAR SYSTEM*

The Inner Planets

The four planets that orbit closest to the sun are called the **inner planets.** They are all relatively small and rocky compared with the other planets.

- **Mercury** is rocky and dry, with lots of craters on its surface. Like our moon, it has no atmosphere.
- **Venus** is about the same size as Earth, and it has a thick atmosphere made mostly of carbon dioxide. It is very hot with temperatures over 800°F.
- **Earth** is the only planet with large quantities of liquid water. As far as we know, it is the only planet in the solar system that supports life.
- **Mars** has a day about as long as an Earth day, and it also has seasons. At the poles, there is dry ice (frozen carbon dioxide) and water ice.

The Outer Planets

The five planets that orbit farthest from the sun are called the **outer planets.**
All of them are made of gas, except Pluto.

- **Jupiter,** the largest planet, is made mostly of hydrogen and helium.
  Its most famous feature is the Great Red Spot, a huge storm system.

- **Saturn,** another gas giant, is best known for its large and complex rings.
  It has the most satellites (moons) of all of the planets.

- **Uranus** appears to be a blue-green color because of methane in its atmosphere.
  It has a ring system, but its rings are irregular and broken.

- **Neptune** is the outermost gas giant. Like Uranus, it is blue-green and has rings.

- **Pluto** is farthest from the sun. It is small and rocky and has a satellite that is
  almost as big as it is.

For more facts about the solar system, see the Science Resources, page 123.

EARTH AND SPACE SCIENCE ■ PRACTICE 5

A. Use the information on pages 48 and 49 to match each planet with its description. Write the letter of the correct planet next to the description.

_____	1.	It has a large and complex ring system.	**a.** Mercury
_____	2.	It is farthest from the sun.	**b.** Venus
_____	3.	It is closest to the sun.	**c.** Earth
_____	4.	Its thick atmosphere is made mainly of carbon dioxide.	**d.** Mars
_____	5.	Its most famous feature is the Great Red Spot.	**e.** Jupiter
_____	6.	Its day is about as long as an Earth day.	**f.** Saturn
_____	7.	It has water and life.	**g.** Uranus
_____	8.	It is the outermost gas giant.	**h.** Neptune
_____	9.	It has an irregular and broken ring system.	**i.** Pluto

B. Use the diagram on page 48 to answer the following question.

10. One planet has an orbit that swings inside the orbit of its neighbor planet, so for
    part of its year it is closer to the sun, and part of its year it is farther from the
    sun than the neighboring planet. Which planet has the irregular orbit?

_____

Answers and explanations start on page 114.

Stars

So far we have explored more than 70 objects in our solar system. Our spacecraft have visited or flown by seven of the eight other planets, some of their satellites, asteroids, and comets. No signs of life have been detected on any of these objects.

When we look outside the solar system, we find billions of stars in the **Milky Way Galaxy.** Which of these stars might have a planet that can support life? Many scientists narrow the search by looking for stars similar to our sun. These are medium-sized, moderately hot stars that have been burning hydrogen and helium for billions of years. Stars like this produce the elements iron, oxygen, and carbon, of which planets and living things are made.

Scientists then look for planets around these stars. Many scientists assume that a planet that supports life would be similar to Earth—rocky, with liquid water. Therefore, the planet must circle the star at the right distance, or water would boil off or freeze. Stars with only gas giant planets are eliminated. They are not likely to support life.

Stars like our sun with rocky planets are most likely to be found midway out from the center of the galaxy. They are not likely to be in areas of the galaxy where stars and dust are very dense.

Galaxies

The Milky Way is our home **galaxy,** a collection of billions of stars, dust, and gas. Galaxies are grouped into three main types:

- **Elliptical galaxies** are round or oval. Their stars are concentrated in the center. They usually consist of old stars and have little gas or dust.
- **Spiral galaxies,** like the Milky Way, are shaped like disks. There is a central bulge, called the nucleus. Spiral galaxies contain young stars mostly located in the spiral arms.
- **Barred spiral galaxies** have a bright "bar" coming out of the nucleus. The spiral arms form off the ends of the bars.

There are also galaxies with no particular shape. These are called irregular galaxies.

Elliptical Galaxy
*Photo Credit: NGC 4636
Optical Image, DSS*

Spiral Galaxy
*Photo Credit: NASA and
The Hubble Heritage Team
(STScI/AURA)*

Barred Spiral Galaxy
*Photo Credit: Allan Sandage
(The Observatories of the
Carnegie Institution of Washington)
and John Bedke
(Computer Sciences Corp. and the
Space Telescope Science Institute)*

TYPES OF GALAXIES

The Universe

The **universe** consists of all matter, energy, and space that exists. Much of the matter in the universe is grouped into galaxies. The galaxies are thought to be moving away from each other. That would mean the universe is expanding.

The universe is thought to have begun when a compact, tiny ball of matter and energy suddenly expanded. This theory is known as the **Big Bang.**

EARTH AND SPACE SCIENCE ▪ PRACTICE 6

A. Based on the information on pages 50 and 51, write *True* or *False* next to each statement. Correct any false statements.

_____ **1.** Spacecraft sent from Earth have not found any sign of life in our solar system.

_____ **2.** Scientists are searching for life on other stars.

_____ **3.** The search for life outside the solar system focuses on planets of stars that are similar to our sun.

_____ **4.** Planets and life forms are made mostly of hydrogen and helium.

_____ **5.** Planets that support life are most likely to exist around stars near the dense center of a galaxy.

_____ **6.** Many scientists think that a planet with liquid water is more likely to support life than a planet without liquid water.

_____ **7.** Galaxies with no particular shape are called irregular galaxies.

B. Use information on page 50 and the diagram of the types of galaxies to answer the following questions.

8. Compare and contrast spiral and barred spiral galaxies.

_____

_____

_____

9. What type of galaxy is the Milky Way Galaxy?

_____

Answers and explanations begin on page 114.

Apply Science Information

When you **apply** information to a specific situation, you are taking what you already know and using it in a new or particular context. For example, suppose you know that Florida has a warm climate. You would use, or apply, this knowledge when packing for a trip to Florida. You would pack summer clothing, not a heavy jacket and warm hat.

On the GED Science Test, some questions will ask you to apply general scientific principles or information to particular situations. For example, you might have to use general information about nutrition to answer a question about a specific food.

When you answer application questions, you use general information that you know or are given, and you apply it to a specific situation.

EXAMPLE

Here is U.S. Geological Survey advice on planning for an earthquake.

- Make sure all family members know what to do wherever they are during an earthquake. Set a place where you can meet afterward.
- Find out the earthquake plans of your children's day-care center or schools.
- Keep emergency food, liquids, and walking shoes at work and home.
- Learn how to turn off gas, water, and electric utilities.
- Find out where the closest police and fire stations and emergency medical centers are.

Mayra is telling her six-year-old son what to do if he's at school during an earthquake. Based on the information above, what advice should she give her son?

    a. Come straight home from school and wait for Mayra in their apartment.

    b. Follow his teacher's directions and wait for Mayra or his father to pick him up afterward.

If you answered *b,* you are correct. A six-year-old should stay with a responsible adult until his parents can meet him. Mayra's son is too young to do anything else.

THINKING STRATEGY: In this case, you had to figure out which of the earthquake planning instructions would apply to a six-year-old. Think: What are the general instructions? How would I use them with a young child?

Now let's look at application questions similar to those on the GED Science Test.

Sample GED Question

The Department of Energy tracks the use of solar energy, a renewable energy resource. The results of this monitoring are shown in the graph below.

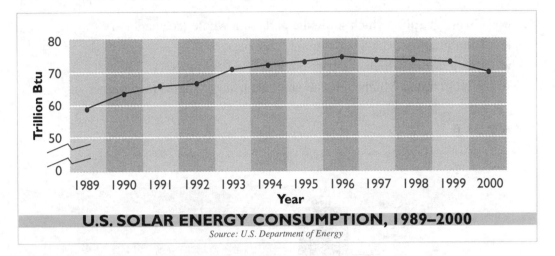

U.S. SOLAR ENERGY CONSUMPTION, 1989–2000

Source: U.S. Department of Energy

You would find the information in this graph most useful if you were

(1) moving to an area with high rainfall

(2) moving to an area with little rainfall

(3) installing a solar collector on your roof

(4) selling solar energy equipment

(5) working at a coal-burning power plant

THINKING STRATEGY: First find the main idea of the graph. Read the title and labels. Look at the trend line. Then ask, in what situation would this information be helpful?

The correct answer is **(4) selling solar energy equipment.** A person selling solar equipment would be interested in how much solar energy is used each year. He or she would also be interested in whether demand has increased, decreased, or stayed steady.

GED THINKING SKILL PRACTICE ■ APPLICATION

Over time, wind-blown sand can sculpt large rocks into strange shapes. Sand is carried along near the ground by the wind. It blows against the lower portion of rock and wears it away. It leaves the upper part intact. Eventually the rock looks a little like a large mushroom.

The action of wind-blown sand is similar to

(1) creating a breeze with a small fan

(2) generating electricity with a windmill

(3) waves depositing sand on a beach

(4) building a sand castle of wet sand

(5) sandblasting a building to clean it

Answers and explanations start on page 114.

Understand Graphs

On the GED Science Test, you will answer questions based on graphs. **Graphs** show numeric information in a visual way so that it is easy to understand. Graphs include:

- **Circle graphs,** which show the parts of a whole (see page 44)
- **Line graphs,** which show trends (see page 53)
- **Bar graphs,** which compare amounts (see below)
- **Pictographs,** which compare amounts using symbols (see page 55)

E X A M P L E

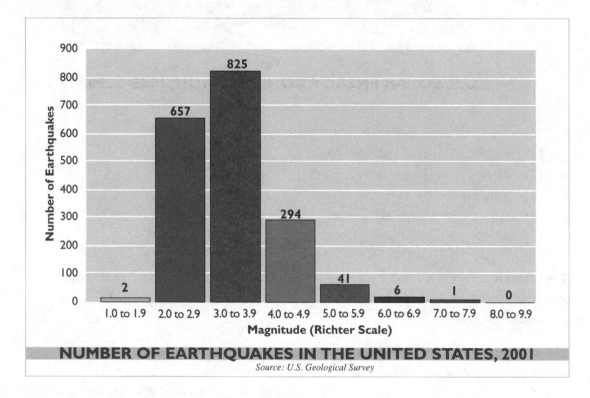

NUMBER OF EARTHQUAKES IN THE UNITED STATES, 2001

Source: U.S. Geological Survey

In 2001, there were more earthquakes of magnitude _____ than of any other magnitude.

Did you answer, *3.0 to 3.9?* If so, you are right. The graph's title indicates that it shows only 2001 earthquake figures. The label along the bottom axis shows magnitude according to the Richter scale. The tallest bar represents the greatest number of earthquakes. In a bar graph, the tallest bar shows the greatest amount; the shortest, the least amount.

When you are asked a question about a graph, first figure out the main idea. See what type of graph it is. Read the title and labels. Finally, locate the detail you are being asked about.

Now let's look at graph questions similar to those on the GED Science Test.

GED Graph Practice

Los Angeles County monitors air pollution levels in many locations. The graph shows the number of days in one year that there was more ozone, a pollutant, than allowed by the state standard.

In 1999, which part of L.A. County had the healthiest air in terms of ozone levels?

(1) Central Los Angeles

(2) Northwest Coastal L.A. County

(3) East San Fernando Valley

(4) East San Gabriel Valley 2

(5) South Central L.A. County

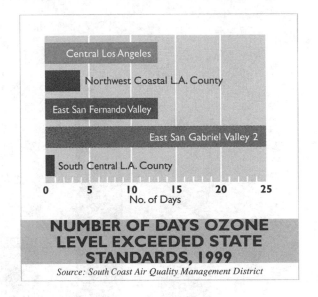

NUMBER OF DAYS OZONE LEVEL EXCEEDED STATE STANDARDS, 1999

Source: South Coast Air Quality Management District

THINKING STRATEGY:

First read the short passage and examine the graph carefully. Figure out what the bars show about ozone levels. Infer how this is related to health.

The correct answer is (5) **South Central L.A. County.** High ozone levels are not healthy. Thus, the healthiest place must be the one with the fewest days—the shortest bar.

GED GRAPHIC SKILL PRACTICE ▪ USING GRAPHS

Questions 1 and 2 refer to the pictograph on the right.

1. How many times has Trident erupted?

   (1) 1

   (2) 3

   (3) 5

   (4) 10

   (5) 15

2. Which volcano erupted 7 times?

   (1) Kiska

   (2) Makushin

   (3) Redoubt

   (4) Trident

   (5) Veniaminof

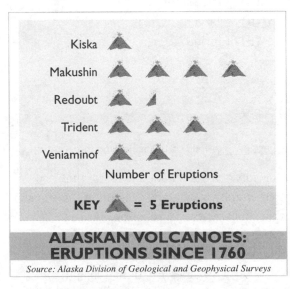

Number of Eruptions

KEY = 5 Eruptions

ALASKAN VOLCANOES: ERUPTIONS SINCE 1760

Source: Alaska Division of Geological and Geophysical Surveys

Answers and explanations start on page 114.

GED Review: Earth and Space Science

Choose the <u>one best answer</u> to the questions below.

<u>Questions 1 and 2</u> refer to the following bar graph.

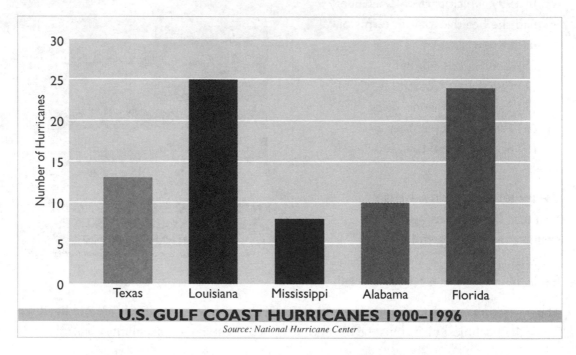

1. Which state had ten hurricanes along its Gulf Coast from 1900 to 1996?
(1) Texas
(2) Louisiana
(3) Mississippi
(4) Alabama
(5) Florida

2. Simone is planning to move to the Gulf Coast for early retirement. If she wants to lower the risk of hurricane damage to her home, to which state should she move?
(1) Texas
(2) Louisiana
(3) Mississippi
(4) Alabama
(5) Florida

<u>Question 3</u> refers to the following information.

Our sun is a star made mostly of the elements hydrogen and helium. In the sun's core, nuclear fusion reactions turn hydrogen into helium. These reactions release huge amounts of energy. Some of this energy reaches Earth as sunlight. Almost all of the energy on Earth comes from sunlight.

3. On Earth, the sun's energy is captured for use by all living things through the process of
(1) respiration
(2) photosynthesis
(3) nuclear fusion
(4) burning fossil fuels
(5) generating electricity

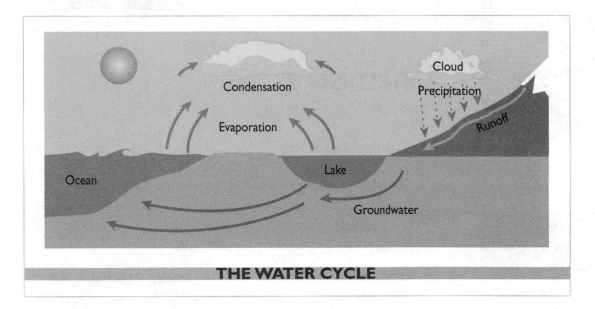

THE WATER CYCLE

4. The diagram shows that water cycles from the atmosphere to the surface and back again. It falls to Earth in the form of
   (1) precipitation
   (2) evaporation
   (3) runoff
   (4) groundwater
   (5) water vapor

5. Humans play a role in the water cycle by
   (1) drinking liquids and eating food
   (2) drinking liquids, excreting urine, and exhaling water vapor
   (3) drinking liquids and passing solid waste
   (4) inhaling oxygen and exhaling water vapor
   (5) inhaling oxygen and exhaling carbon dioxide

Question 6 refers to this circle graph.

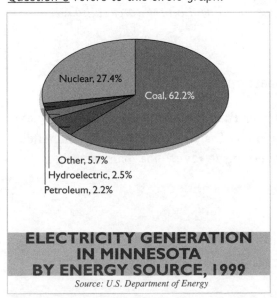

ELECTRICITY GENERATION IN MINNESOTA BY ENERGY SOURCE, 1999
Source: U.S. Department of Energy

6. About one-quarter of the electricity in Minnesota is generated using
   (1) coal
   (2) petroleum
   (3) nuclear power
   (4) hydroelectric power
   (5) other sources of energy

Answers and explanations begin on page 115.

Chemistry

SCIENCE SKILLS

- Understand the nature and states of matter
- Define and identify elements, mixtures, and compounds
- Recognize different types of chemical reactions

THINKING SKILL

- Analyze science information

GRAPHIC SKILL

- Understand charts and tables

GED REVIEW

1. Think About the Topic

The program that you are about to watch is about *Chemistry*. It will introduce some of the basic ideas in chemistry—atoms and elements, states of matter, chemical reactions, and the periodic table.

In this video, you will see chemistry teachers, chemists, and others discussing the basics of chemistry. They will also show you how people use chemistry in everyday life. How do you think chemistry plays a role in your life?

2. Prepare to Watch the Video

Matter and its different states are important topics covered in the program. Matter can exist as a solid, a liquid, or a gas. On the lines below, write how water can exist in different states.

_____

_____

_____

You may have written something similar to: *Freezing water (a liquid) can make ice (a solid). Boiling water (a liquid) creates steam (a gas).*

3. Preview the Questions

Read the questions under *Think About the Program* on the next page, and keep them in mind as you watch the program. After you watch, use the questions to review the main ideas in the program.

4. Study the Vocabulary

Review the terms to the right. Understanding the meaning of key chemistry vocabulary will help you understand the video and the rest of this lesson.

WATCH THE PROGRAM

As you watch the program, pay special attention to the host who introduces or summarizes major chemistry ideas that you need to learn about. The host will also give you important information about the GED Science Test.

AFTER YOU WATCH

1. Think About the Program

Why is it important to study chemistry?

What are some of the differences between elements, mixtures, and compounds?

How can a basic understanding of chemistry help you with everyday tasks, such as cooking?

How can the periodic table be used to find similarities among groups of elements?

2. Make the Connection

The program discussed the effect that chemicals can have on the environment. Look in newspapers and magazines to find articles about pollution. Read carefully and list problems caused by the misuse of chemicals in the environment.

atom—the smallest unit of matter that cannot be broken down into particles by chemical means

chemical reaction—the changing of one set of substances into another set of substances with different properties

compound—a substance in which two or more elements are chemically combined

density—the amount of matter that something contains per unit of volume

element—a substance in which all the atoms are the same kind

mass—the amount of matter that something contains

matter—anything that has mass and takes up space

mixture—a combination of two or more substances that keep their original properties

periodic table—a table in which elements are arranged in order of atomic number and properties

solution—a mixture in which tiny particles of two or more substances mix evenly

state of matter—whether matter is in the form of a solid, liquid, or gas

"Chemistry is the study of matter—what it is and how it changes."

Matter: What Everything Is Made Of

Matter takes up space and has mass. Iron, orange juice, and air are all matter. They take up space and have mass. **Mass** is the amount of matter a substance has. Anything that has mass can be weighed. Weight is a measure of the pull of gravity on an object. An object's weight depends on where you weigh it. For example, say you have a barbell that weighs 100 pounds on Earth. If you took that barbell to the moon, it would weigh only 17 pounds. That's because the pull of gravity on the moon is weaker than the pull of gravity on Earth.

The Properties of Matter

All matter has physical properties that our senses can detect. These include:

- *Color.* By looking, you see that a baseball is white and a tennis ball is yellow. Some matter is colorless.
- *Odor.* By smelling, you can tell whether someone is wearing perfume or not. Some matter has no odor, however.
- *Taste.* By tasting, you can tell the difference between a cola drink and ginger ale. But some matter is tasteless.
- *Hardness.* By touching, you can tell how hard matter is. It may be as hard as a diamond or as soft as a pillow.

Another important physical property of matter is **density.** Density is the amount of mass in a specific volume of matter. The more mass per unit of volume, the denser an object is. For example, one cubic foot of solid steel is denser than one cubic foot of plastic foam. The steel has more mass, even though it occupies the same amount of space as the plastic foam.

The Structure of the Atom

As you saw on the video, atoms are the building blocks of matter. **Atoms** are the smallest unit of matter that cannot be broken down into particles by chemical means.

Atoms have a dense nucleus. In the nucleus are two types of particles: protons and neutrons. **Protons** have a positive (+) charge. **Neutrons** are neutral; they have no electric charge. Surrounding the nucleus are **electrons.** These have much less mass than protons and neutrons. They also have a negative (−) electric charge. The diagram on the next page shows a carbon atom.

Notice that the carbon atom has 6 protons, 6 neutrons, and 6 electrons. There is the same number of protons as electrons in the atom. That means the positive charge (+6) and the negative charge (−6) cancel one another out. Thus, an atom of carbon is electrically neutral.

Notice also that the electrons circle the nucleus at two different energy levels. At the first level are two electrons. At the second level are four electrons.

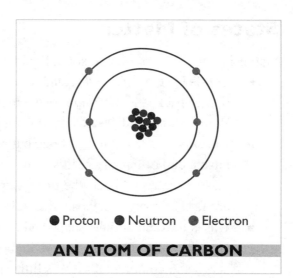

● Proton ● Neutron ● Electron

AN ATOM OF CARBON

CHEMISTRY ▪ PRACTICE 1

A. Use the information and diagram on pages 60 and 61 to complete the following sentences.

1. All matter has mass and takes up _____.

2. You can tell how much mass an object has on Earth by _____ it.

3. On the moon, objects weigh less than they do on Earth because the moon's _____ is weaker than that of Earth.

4. All matter has _____ that can be observed by our senses.

5. _____ is the amount of mass an object has per unit of volume.

6. A cubic foot of steel is _____ than a cubic foot of air.

7. A carbon atom has six _____, six _____, and six _____.

8. Because it has an equal number of positively charged and negatively charged particles, a carbon atom is electrically _____.

B. Use the information on pages 60 and 61 to answer the following question.

9. Choose an object you see or use every day and describe its physical properties, including color, density, texture, hardness, odor, and possibly taste (if appropriate).

_____

_____

Answers and explanations start on page 115.

States of Matter

Matter is found in three states: solid, liquid, and gas.

- **Solids** have a definite size and a definite shape. The atoms in a solid are very tightly packed and hardly move. Often they form regular patterns, like crystals. Wood, stone, and plastic are all examples of solids.

- **Liquids** have a definite volume but no definite shape. This means that if you have a cup of lemonade, it is a cup of liquid whether it's in a gallon carton, in a glass, or in a puddle on the floor. Liquids can change shape because their atoms are farther apart than those of a solid. They can move more easily.

- **Gases** have no definite shape and no definite volume. The atoms in a gas are very far apart. They are always in motion. A gas will spread out if there's a lot of space and compress if there is little space. For example, oxygen is spread out through the air, but it is compressed in an oxygen tank.

Changes of State

Matter can change from one state to another through the addition of heat energy. When you add heat to a substance, its atoms gain energy and move more freely. For example, when you add heat to a solid like ice, the energy causes the atoms to move around much more. The ice melts, becoming water.

Matter can also change from one state to another through the removal of heat energy. When you take heat out of a substance, its atoms lose energy and their motion slows. For example, when you place water in a freezer, it loses energy and freezes.

The table below summarizes the changes of state of matter.

Process	What Happens	Example
ADDING HEAT ENERGY		
Melting	A solid becomes a liquid.	Ice melts and turns into water.
Boiling	A liquid becomes a gas.	Water boils and turns into steam (water vapor).
REMOVING HEAT ENERGY		
Condensing	A gas becomes a liquid.	Water vapor condenses and turns into water.
Freezing	A liquid becomes a solid.	Water freezes and becomes ice.

CHANGES OF STATE

SCIENCE SKILLS

A. Use the information and table on page 62 to answer the following questions. Place the letter of the correct answer in the space provided.

_____ **1.** This substance has atoms that are very far apart. They are in constant, high-energy motion. The substance spreads out to fill any space, and it has no definite shape. This substance is a
  a. solid   **b.** liquid   **c.** gas

_____ **2.** This substance has a definite shape and a definite volume. Its atoms hardly move, because they are tightly packed in a regular pattern. This substance is a
  a. solid   **b.** liquid   **c.** gas

_____ **3.** This substance has a definite volume but no definite shape. Its atoms can spread out and move quite easily. This substance is a
  a. solid   **b.** liquid   **c.** gas

_____ **4.** According to the chart, which two processes involve adding heat energy to matter?
  a. melting and boiling   **b.** boiling and condensing
  c. condensing and freezing   **d.** freezing and melting

_____ **5.** According to the chart, which two processes involve removing heat energy from matter?
  a. melting and boiling   **b.** boiling and condensing
  c. condensing and freezing   **d.** freezing and melting

B. Use the information and table on page 62 to answer the following questions.

6. Compare and contrast melting and freezing. How are they similar? How are they different?

_____

_____

_____

7. Liquids can turn into gas by boiling or by evaporation. In evaporation, atoms at the surface of the liquid absorb heat energy and escape from the liquid, turning into a gas. What is a common example of evaporation in everyday life?

_____

_____

Answers and explanations start on page 116.

Elements, Compounds, and Mixtures

Elements

An **element** is a substance in which all the atoms are of the same kind. On page 61, there is a diagram of an atom of one element—carbon. Carbon is an element that makes up part of carbon dioxide gas in the atmosphere. It is the main substance in coal, graphite, and diamonds. It is also found in all living things. You are probably familiar with other elements. They include metals like gold, copper, and silver. They include nonmetals like silicon, arsenic, and iodine. They include oxygen, nitrogen, helium, and neon, which are gases, and bromine and mercury, which are liquids at temperatures generally found on Earth.

More than 110 elements have been discovered so far. Of these, more than 92 have been found naturally on Earth. The remaining elements have been made in laboratories. Chemists arrange all the elements in order of atomic number in the periodic table. (The periodic table appears in the Science Resources section on pages 124–125.)

Each element has its own name, chemical symbol, and atomic number. The **chemical symbol** is an abbreviation that consists of one or more letters. The first letter is always capitalized. For example, C is the chemical symbol for carbon. Ne is the chemical symbol for neon.

The **atomic number** is the number of protons in the nucleus of one atom of the element. The atomic number of hydrogen is 1 because hydrogen has only one proton in its atomic nucleus. The atomic number of gold is 79 because gold has 79 protons in its atomic nucleus. Look at the diagram of carbon on page 61. What is the atomic number of carbon?*

Each element occupies one block of the periodic table. There are many formats for the blocks of the periodic table. A common format is shown on the right. In this example, the name of the element is calcium. Its chemical symbol is Ca. And its atomic number is 20, meaning that there are 20 protons in the nucleus of a calcium atom. Look for Ca in the periodic table on pages 124–125.

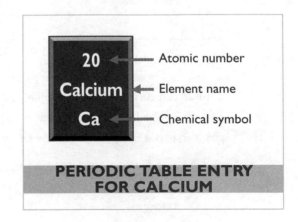

PERIODIC TABLE ENTRY
FOR CALCIUM

*The atomic number of carbon is 6.

Properties of Elements

Elements have physical properties like color, odor, and taste. They also have chemical properties. Some elements combine easily with other elements in chemical reactions. In the video, you saw the elements magnesium, a solid metal, and oxygen, a gaseous nonmetal, combine. Both magnesium and oxygen combine easily with other elements. Elements that combine easily with other elements are reactive.

Other elements do not combine easily with other elements. For example, gold does not combine easily with other elements. Neither do nitrogen, helium, and argon. These elements are unreactive.

CHEMISTRY ▪ PRACTICE 3

A. Based on the information on pages 64 and 65, write *True* or *False* next to each statement.

_____ **1.** An element is a substance made up of just one kind of atom.
_____ **2.** Every element we have discovered so far occurs naturally on Earth.
_____ **3.** Scientists have made some elements in the laboratory.
_____ **4.** Carbon is an element that is found in all living things.
_____ **5.** All elements react easily with other elements.

B. Use the following information about the elements sodium and chlorine to make a periodic table entry for each. Follow the format of the calcium entry on page 64.

6. Element name: sodium
Number of protons: 11
Chemical symbol: Na

7. Element name: chlorine
Number of protons: 17
Chemical symbol: Cl

Answers and explanations start on page 116.

Compounds

A second type of substance is called a compound. **Compounds** form when two or more elements chemically combine in a definite proportion. For example, water is a compound consisting of the elements hydrogen and oxygen. The proportion of hydrogen to oxygen in water is always 2:1. One **molecule,** or unit, of the compound water can be represented by the chemical formula H_2O. This means that one molecule of water consists of 2 hydrogen atoms and 1 oxygen atom.

The properties of a compound differ from the properties of the elements of which it is made. For example, water—a liquid compound (at room temperature)—is made of the elements hydrogen and oxygen, which are gases at room temperature.

There are about 10 million known compounds. In addition to water, many other familiar substances are compounds:

- Salt, or sodium chloride (NaCl), is a compound made of sodium (Na) and chlorine (Cl).
- Quartz, or silicon oxide (SiO), is a compound composed of silicon (Si) and oxygen (O).
- Bottled gas, or propane (C_3H_8), is a compound composed of carbon (C) and hydrogen (H).

Mixtures

A third type of substance is called a mixture. **Mixtures** are combinations of two or more substances that keep their original properties. The substances that form mixtures do not combine chemically. Instead, they mix physically. Soil, salad dressing, and perfumed air are all mixtures. The substances in a mixture can combine in any proportion.

A special type of mixture is called a **solution.** In a solution, very tiny particles of two or more substances mix evenly. An example of a liquid solution is salt water. In salt water, the compound salt is the **solute,** the substance that dissolves. The compound water is the **solvent,** the substance in which the solute dissolves. Not all solutions are liquid. Solutions can be solid or gaseous, too. For example, steel is a solid solution of carbon in iron. Air is a gaseous solution of nitrogen, oxygen, and other trace gases.

The ingredients of a solution, like those of any mixture, can be separated out. For example, if you heat salt water, the water will evaporate, leaving the salt behind. This is the principle behind extracting salt from seawater. Seawater sits in shallow ponds and evaporates in the heat of the sun. Blocks of sea salt are left behind.

Compounds and Mixtures Compared

The following table summarizes the characteristics of compounds and mixtures.

	Compound	Mixture
Composition	Made up of elements in definite proportions	Made up of substances in any proportion
Properties	The properties of the elements that make up the compound are different from the properties of the resulting compound.	The properties of the substances that make up the mixture are retained.
Examples	Water, salt, hydrochloric acid, ammonia, methane	Sand, gravel, mayonnaise, milk, air, tea, sugar water

CHARACTERISTICS OF COMPOUNDS AND MIXTURES

CHEMISTRY ■ PRACTICE 4

A. Based on the information on pages 66 and 67, write _True_ or _False_ next to each statement.

_____ 1. A compound is a chemical combination of two or more elements.

_____ 2. Common compounds include water, salt, and propane.

_____ 3. In chemistry, a compound can be represented by a chemical formula.

_____ 4. The chemical formula for water is HO.

_____ 5. Mixtures consist of two or more substances in a definite proportion.

_____ 6. Once a mixture forms, it is impossible to separate out its substances.

B. Use information on page 66 and the table on page 67 to answer this question.

7. Compare and contrast compounds and mixtures. How are they similar? How are they different?

_____

_____

_____

_____

Answers and explanations start on page 116.

"You don't have to be in a lab to cause a chemical reaction. You cause chemical reactions every day."

Chemical Reactions

On the video, chemistry teacher Charlotte Ray adds sulfuric acid to sugar. When these substances react, a "tower" of black carbon arises in the beaker, and water escapes into the air. By combining a yellow liquid and a white solid, Ms. Ray produces a black solid and a colorless gas that condenses as steam in the air.

What Is a Chemical Reaction?

Ms. Ray's experiment illustrates one of the key characteristics of **chemical reactions:** one set of substances, called the **reactants,** is changed into another set of substances, called the **products.** The change takes place because the atoms of the reacting substances rearrange themselves and form new substances.

Reactants and products can be either elements or compounds. In Ms. Ray's experiment, the reactants, sulfuric acid and sugar, are both compounds. One product, carbon, is an element, and the other product, water vapor, is a compound. Thus, the products of a reaction usually have very different chemical properties than the reactants have.

A second key characteristic of chemical reactions is that matter is neither created nor destroyed. It simply changes from one form to another. Therefore, the total mass of the reactants equals the total mass of the products. If you found the mass of the reactants before the reaction and the mass of the products after the reaction, the masses would be the same. This principle is called the **Law of Conservation of Matter.**

Energy and Chemical Reactions

Before a chemical reaction can occur, there must be enough energy to allow the reactants to change. The energy needed to get a chemical reaction started is called the **activation energy.**

In some chemical reactions, energy is released when the reactants change. These reactions are called **exothermic reactions.** Burning is an example of an exothermic reaction.

In other chemical reactions, energy is absorbed when the reactants change. These reactions are called **endothermic reactions.** Hard-boiling an egg is an endothermic reaction. The table at the top of the next page summarizes exothermic and endothermic reactions.

Type of Reaction	What Happens	Examples
Exothermic reaction	Energy is released in the form of heat and light.	Combustion (burning), cellular respiration
Endothermic reaction	Energy is absorbed.	Caramelization (making caramel), photosynthesis

EXOTHERMIC AND ENDOTHERMIC REACTIONS

CHEMISTRY ▪ PRACTICE 5

A. Use the information on page 68 to match each term with its definition. Write the letter of the term in the space provided.

_____ **1.** a reaction in which energy is released

_____ **2.** the substances that result from a chemical reaction

_____ **3.** the substances that change during a chemical reaction

_____ **4.** the change of one set of substances into another set of substances through the rearrangement of atoms

_____ **5.** a reaction in which energy is absorbed

_____ **6.** the energy needed to start a chemical reaction

a. chemical reaction
b. endothermic reaction
c. activation energy
d. products
e. reactants
f. exothermic reaction

B. Use the information on pages 68 and 69 to answer the following questions.

7. Explain what is meant by this sentence: "In a chemical reaction, matter is neither created nor destroyed."

_____

_____

8. How do endothermic and exothermic reactions differ from one another?

_____

9. Is energy absorbed or released during the process of photosynthesis in green plants? Explain.

_____

_____

Answers and explanations start on page 116.

Everyday Chemistry

Kitchen Chemistry

In the video, Harold McGee demonstrates some kitchen chemistry. He combines an acid and a base to produce carbon dioxide bubbles. The bubbles are trapped in pancake batter and lighten the pancakes.

If you have ever baked a cake or bread, you know that many baked goods are made with a **leavening agent.** The leavening agent lightens the dough or batter. In some recipes, the leavening agent is baking powder. Baking powder contains baking soda— a base—and tartaric acid. In these recipes, the baking soda and tartaric acid react when a liquid is added, producing carbon dioxide, water, and a salt. Other recipes call for baking soda. These recipes must also include an acid, such as the acid found in buttermilk or sour milk, in order to produce carbon dioxide.

Most bread and pizza doughs use yeast as the leavening agent. Yeasts are living organisms that need food and warmth. In baking, yeast gets food from sugar. If you put the dough in a warm place, the yeast becomes active. It breaks down the sugar, producing carbon dioxide bubbles and alcohol. This process is known as **fermentation.** As you can see in the diagram, dough can rise to several times its volume because of fermentation.

After mixing   After two hours

FERMENTATION IN A YEAST DOUGH

Source: Joy of Cooking 1964, p. 554

First-Aid Chemistry

Have you ever used hydrogen peroxide to wash a cut? If so, you know that when you put it on a cut, it foams. If you put it on unbroken skin, however, it doesn't foam.

Hydrogen peroxide is a compound made up of hydrogen and oxygen. Hydrogen peroxide foams on a cut or scrape because it reacts with an enzyme in blood. When the enzyme, called catalase, comes in contact with hydrogen peroxide, it breaks the hydrogen peroxide down into water and oxygen. The bubbles in the foam are oxygen gas.

Soap and Detergent Chemistry

Soap is a compound formed from an alkali metal and a fatty acid. Detergents are compounds made of the sodium salts of acids that contain sulfur. Despite their different chemical compositions, both soap and detergent work in a similar way. They both make grease dissolve in water. Since grease holds dirt on the skin and clothes, removing it also removes the dirt. That's why parents commonly tell their children, "Wash with soap!" They know that water alone does not clean well.

A. Use the information on pages 70 and 71 to complete these sentences.

1. A leavening agent lightens dough or batter by producing _____ bubbles.

2. In many baking recipes, the leavening agent is _____, a combination of baking soda and tartaric acid.

3. If the leavening agent is baking soda, then the recipe must include a(n) _____ in order to produce bubbles.

4. A living organism called _____ is the leavening agent in many breads.

5. When hydrogen peroxide is applied to a cut or scrape, it breaks down into _____ and _____.

6. Soaps and detergents remove dirt by dissolving _____ in _____.

B. Use the information on pages 70 and 71 to answer the following questions.

7. Both matzos and communion wafers are unleavened breads. What ingredients are specifically NOT in unleavened breads?

   _____

8. Why doesn't hydrogen peroxide foam when you pour it on your skin?

   _____

9. Describe one difference between soap and detergent.

   _____

   _____

Answers and explanations begin on page 117.

Analyze Science Information

When you **analyze** information, you break it down into parts. You then examine the relationships among the parts.

For example, suppose you woke up one winter morning and heard on the radio that schools would be closed that day. You would consider the season, the hour, the source of your information, and the fact that schools will be closed. From these facts, you might conclude that there was significant snowfall during the night.

On the GED Science Test, some questions will ask you to draw conclusions based on evidence. Some will ask you to compare and contrast two items. Some will ask you to determine cause and effect. All of these thinking skills involve analysis.

When you answer analysis questions, you break down information into parts and examine the relationships among those parts. You may have to draw a logical conclusion based on the evidence in the passage or graphic. You may have to determine the cause or effect of an event. Or you may have to figure out how two items, processes, or ideas are alike or how they are different.

EXAMPLE

> Teflon, a type of plastic, has been used to line frying pans for years. Its slippery surface repels grease and food so they don't stick to the pan. Recently, chemists and textile designers teamed up to use Teflon on fabrics. Coating fibers with Teflon protects fabrics from greasy stains. You can spill food on a Teflon-treated tie or shirt and not have to take it to the dry cleaner's.

What causes Teflon to repel grease and food?

a. It acts like a detergent to remove grease and food from pans and fabrics.
b. It has a slippery surface so grease and food don't stick to it.

If you answered *b,* you are correct. According to the passage, the slippery surface of Teflon causes grease and food to be repelled.

THINKING STRATEGY: In this case, you were being asked a cause-effect question. The question gave you the effect: Teflon's ability to repel grease and food. It asked you for the cause. What characteristic of Teflon causes it to repel matter? To find the answer, you need to reread the passage carefully, looking for the answer to the question.

Now let's look at other analysis questions similar to those you will see on the GED.

Sample GED Question

	Low-Alloy Steel	**High-Alloy Steel**
Composition	Iron, carbon, and less than 5 percent other metals	Iron, carbon, and more than 5 percent other metals, including chromium
Characteristics	Extremely strong; rusts easily	"Stainless"; does not rust easily
Uses	Ships, cars, bridges, machine parts	Eating and cooking utensils, bank vaults

TYPES OF STEEL

In terms of their composition, what do low-alloy and high-alloy steels have in common?

(1) Both contain chromium.

(2) Both contain iron and carbon.

(3) Both contain more than 5 percent of metals other than iron.

(4) Both contain less than 5 percent of metals other than iron.

(5) Both are rust-resistant.

THINKING STRATEGY: First find the row that gives information about the composition of steel. Then compare the two types of steel to see how they are alike.

The correct answer is **(2) Both contain iron and carbon.** According to the chart, both low-alloy steel and high-alloy steel contain iron and carbon.

GED THINKING SKILL PRACTICE ▪ ANALYSIS

Marinating meat and fish before grilling improves their taste and texture. According to scientists, it may also decrease the production of heterocyclic amines. These are compounds known to cause cancer. The amines give food grilled at high temperature that "grilled" flavor. The compounds have been linked to breast and colon cancer in laboratory animals.

From this information, you can conclude that marinating improves the flavor of grilled food and may also

(1) cause cancer in human beings

(2) cause the food to burn on the grill

(3) increase the production of cancer-causing compounds

(4) cause cancer in laboratory animals

(5) help prevent some cancers

Answers and explanations start on page 117.

Understand Charts and Tables

On the GED Science Test, you will answer questions based on **charts** and **tables.** Both show information arranged in vertical columns and horizontal rows. The resulting grid makes it easy to find specific information. When you look at a chart or table, read the title first. That will tell you the topic. Then read the names of the columns and rows to see what type of information each contains. Finally, look at the whole graphic.

EXAMPLE

Common Name	Chemical Name	Chemical Formula	Structural Formula
Water	Hydrogen oxide	H_2O	
Natural gas	Methane	CH_4	
Grain alcohol	Ethanol	C_2H_5OH	

WAYS TO REPRESENT COMMON COMPOUNDS

1. What is the topic of this table?

_____

Did you answer, *different ways to show chemical compounds?* If so, that is correct. The topic of a table is usually found in the table title.

2. What is the chemical formula for natural gas?

_____

The right answer is CH_4. To find the answer, look in the first column, "Common Name," for natural gas. Then read across the natural gas row until you find the chemical formula in the "Chemical Formula" column. There you will find the answer.

Now let's look at table questions similar to those you will see on the GED Science Test.

GED Table Practice

In the periodic table, elements are arranged in rows in order of increasing atomic number. The vertical columns are called groups or families. Elements in the same group have similar properties.

Which of the following elements is most likely to have properties similar to those of phosphorus?

(1) zinc

(2) silicon

(3) arsenic

(4) sulfur

(5) argon

						2 Helium He
5 Boron B	6 Carbon C	7 Nitrogen N	8 Oxygen O	9 Fluorine F	10 Neon Ne
13 Aluminum Al	14 Silicon Si	15 Phosphorus P	16 Sulfur S	17 Chlorine Cl	18 Argon Ar
30 Zinc Zn	31 Gallium Ga	32 Germanium Ge	33 Arsenic As	34 Selenium Se	35 Bromine Br	36 Krypton Kr

■ **Transition Metals**   ■ **Nonmetals**

PORTION OF THE PERIODIC TABLE

THINKING STRATEGY: First read the short passage and the question. Examine the table. Look for phosphorus. Then look for elements in the same group as phosphorus.

The correct answer is **(3) arsenic.** Of the choices, only arsenic is in the same group as phosphorus. Therefore, arsenic must have properties similar to those of phosphorus.

GED GRAPHIC SKILL PRACTICE ▪ USING TABLES

Question 1 refers to the table below.

Energy Level	Maximum Number of Electrons
1	2
2	8
3	18
4	32
5	50 (theoretically)

ARRANGEMENT OF ELECTRONS IN AN ATOM

Question 2 refers to the table below.

Substance	Boiling Point (°C)
Water	100
Oxygen	−183
Nitrogen	−196
Mercury	357
Iron	2,750

BOILING POINTS OF COMMON SUBSTANCES

1. In an atom, which energy level holds a maximum of 8 electrons?

    (1) energy level 1

    (2) energy level 2

    (3) energy level 3

    (4) energy level 4

    (5) energy level 5

2. Which has the lowest boiling point?

    (1) water

    (2) oxygen

    (3) nitrogen

    (4) mercury

    (5) iron

Answers and explanations start on page 117.

GED Review: Chemistry

Choose the <u>one best answer</u> to the questions below.

<u>Questions 1 and 2</u> refer to the following table.

Use	Amount of Plant Food	Amount of Water
Outdoor plants	1 tablespoon	1 gallon
Houseplants—feed every four weeks	1 teaspoon	1 gallon
Houseplants—feed with every watering	$\frac{1}{4}$ teaspoon	1 gallon
Seeds started outdoors	1 tablespoon	1 gallon
Seeds started indoors	1 teaspoon	1 gallon

SOLUTIONS OF PLANT FOOD FOR VARIOUS USES

1. If you are feeding houseplants every four weeks, how much plant food should you dissolve in one gallon of water?
    (1) $\frac{1}{4}$ teaspoon
    (2) $\frac{1}{2}$ teaspoon
    (3) 1 teaspoon
    (4) 1 tablespoon
    (5) 1 gallon

2. For which use do you mix the weakest solution of plant food and water?
    (1) outdoor plants
    (2) houseplants fed every four weeks
    (3) houseplants fed with every watering
    (4) seeds started outdoors
    (5) seeds started indoors

3. White table sugar, or sucrose, has the following chemical formula:

    $$C_{12}H_{22}O_{11}$$

    Sucrose can be broken down into simple sugars like glucose. The chemical formula for glucose is:

    $$C_6H_{12}O_6$$

    What do sucrose and glucose have in common?
    (1) Both have 12 carbon atoms.
    (2) Both have 12 hydrogen atoms.
    (3) Both have 6 oxygen atoms.
    (4) Both consist of carbon, hydrogen, and oxygen atoms.
    (5) Both are white table sugars.

	Ammonia	Chlorine Bleach	Glass Cleaner
General Warnings	Do not mix with chlorine bleach or other household cleaners. Dangerous fumes result.	Do not mix with ammonia or other household cleaners. Dangerous fumes result.	Contains ammonia. Do not mix with chlorine bleach or other household cleaners. Dangerous fumes result.
If Splashed in Eye	Flush with water for 15 minutes. Call Poison Control Center.	Flush with water for 15 minutes. Call Poison Control Center.	Flush with water for 15 minutes. Call Poison Control Center.
If Swallowed	Do not induce vomiting. Give 1 or 2 glasses of milk or water. Call Poison Control Center.	Do not induce vomiting. Give 1 or 2 glasses of milk or water. Call Poison Control Center.	Do not induce vomiting. Give 1 or 2 glasses of milk or water. Call Poison Control Center.

CAUTIONS FROM LABELS ON COMMON HOUSEHOLD CLEANERS

4. The first aid for both eye contact and swallowing these cleaners involves water or milk. What is one result of this first aid?
   (1) It weakens the strength of the cleaner.
   (2) It intensifies the action of the cleaner.
   (3) It makes the cleaner harmless.
   (4) It distracts the victim from his or her pain.
   (5) It gives the Poison Control Center time to figure out what to do.

5. According to the table, when ammonia and chlorine bleach are combined, dangerous fumes result. This suggests that which of the following occurred?
   (1) a change of state from solid to liquid
   (2) a change of state from liquid to gas
   (3) a change of state from gas to liquid
   (4) a chemical reaction producing a poisonous gas
   (5) a chemical reaction producing a crystalline solid

6. Most solids melt if enough heat is applied. However, some go directly from a solid to a gaseous state. Examples are dry ice (solid carbon dioxide) and iodine. This process is called sublimation.

   What causes sublimation?
   (1) adding heat
   (2) removing heat
   (3) adding water
   (4) removing water
   (5) adding crystals

Answers and explanations begin on page 117.

Physics

SCIENCE SKILLS

- Learn the relationships between motion, work, and energy
- Understand different types of waves
- Know what magnetism and electricity are and how they interact

THINKING SKILL

- Evaluate science information

GRAPHIC SKILL

- Understand diagrams

GED REVIEW

I. Think About the Topic

The program you are about to watch is on *Physics*. In this program, you will learn basic information about motion, heat, sound, electricity, and magnetism.

In this program, you will meet professionals and physics teachers as they present important science concepts. You will hear a pair of individuals who work at NASA discussing the second law of motion: *force = mass × acceleration*. This means that the amount of force needed to move an object depends on its mass, which is related to its weight. This law explains why launching a rocket into space requires more force than moving a car down the road does.

2. Prepare to Watch the Video

In this video, you will hear the host discuss the laws of thermodynamics. He will mention that heat energy always goes from a hotter item to a cooler item. What are two real-life examples of this law? _____

_____

_____

You may have written something similar to: *If you sit next to a heater on a cold day, you warm up. When you put a frozen dinner in a hot oven, it warms up.*

3. Preview the Questions

Read the questions under *Think About the Program* on the next page, and keep them in mind as you watch the program. After you watch, use the questions to review the main ideas in the program.

4. Study the Vocabulary

Review the terms to the right. Understanding the meaning of key physics vocabulary will help you understand the video and the rest of this lesson.

WATCH THE PROGRAM

As you watch the program, pay special attention to the host who introduces or summarizes major physics ideas that you need to learn about. The host will also give you important information about the GED Science Test.

AFTER YOU WATCH

1. Think About the Program

What makes something a "law" in physics?

Why might it be important to understand the laws of motion?

In what ways have scientists used their knowledge of physics to help people?

How is sound similar to waves in the ocean?

2. Make the Connection

In this video you see a team of workers preparing the sound for a production of the play *Annie*. Think of a time you attended a concert or play. How was the sound? Where was the sound the best? What do you think caused the differences in sound quality?

circuit—the path taken by an electric current

electricity—a form of energy produced by the flow of electric charges through a conductor

electromagnetic spectrum—the range of electromagnetic waves, including visible light

energy—the ability to do work

equilibrium—a state in which all forces are in balance

force—the push or pull exerted on an object that causes the object to accelerate

frequency—the number of waves produced in one second

gravity—the attractive force between objects due to mass

inertia—the tendency for an object to stay at rest or in motion unless acted upon by a force

magnetism—a force produced by charged particles that are moving or spinning

wave—a back-and-forth disturbance that repeats itself

work—the movement that results when a force is applied to an object

> *"There are many different kinds of energy, including some you may not think of as energy at all."*

Motion, Work, and Energy

Physics is the study of the relationship between matter and energy. It includes motion, forces, waves, electricity, and magnetism. As in the video, we'll begin with forces and motion.

Force

A **force** is a push or pull that accelerates (changes the speed or direction of) an object. We are constantly being acted on by one force: **gravity.** Gravity keeps everything on Earth's surface from floating away into space. People also exert force. When you use your muscles to move something, you are exerting force. Force is often measured in newtons or pounds.

In most situations, several forces are acting on an object at the same time. For example, in a tug-of-war between two kids, one kid pulls with a force of 60 pounds. The other kid pulls with a force of 40 pounds in the opposite direction. The resultant force *R* is 20 pounds toward the stronger kid. If two kids are equally matched in a tug-of-war, then they exert the same amount of force. The resultant force in that situation is zero—the individual forces are exerted, but the system is in **equilibrium.**

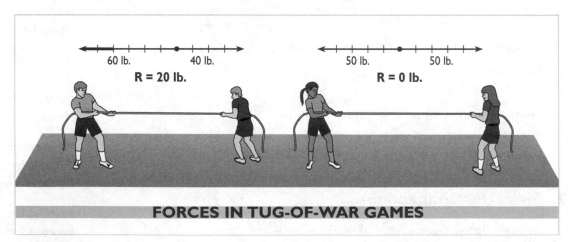

60 lb.   40 lb.   50 lb.   50 lb.
R = 20 lb.   **R = 0 lb.**

FORCES IN TUG-OF-WAR GAMES

Motion

Everything in the universe is in motion. An object in motion has speed, the rate at which it moves, such as 55 miles an hour. It also has direction, such as east. Speed plus direction defines the **velocity** of an object—in this case, 55 miles per hour to the east. In the 1600s, Sir Isaac Newton formulated laws that explain why objects move.

Law	What It Says	Example
First Law	Every object remains in a state of rest or of uniform motion unless acted upon by a force.	A suitcase will stay in place until you lift it.
Second Law	If a force acts on an object, the object accelerates (speeds up) in the direction of the force. The force is equal to the mass of the object times its acceleration.	If you lift a suitcase, it accelerates upward. It takes more force to lift a 40-pound suitcase than to lift a 20-pound suitcase.
Third Law	When two objects interact, the force exerted by the first object on the second object results in an equal and opposite force exerted by the second on the first.	When a 20-pound suitcase is placed on a table, exerting a downward force on the table, the table exerts an equal and opposite upward force on the suitcase.

NEWTON'S LAWS OF MOTION

PHYSICS ■ PRACTICE 1

A. Use the information on pages 80 and 81 to complete these sentences.

1. The push or pull that accelerates an object is called a(n) _____.

2. _____ is a pulling force that pulls objects to the surface of Earth.

3. _____ is the speed at which an object is moving in a particular direction.

4. According to the _____, you need more force to accelerate a heavy truck than a light car.

5. According to the first law of motion, an object in motion in a straight line tends to stay in motion in a straight line unless a _____ acts on it.

6. According to the third law of motion, when object A exerts 20 pounds of force on object B, object B exerts a(n) _____ and _____ force on object A.

B. Use the information in the table above to answer the following question.

7. The Voyager spacecraft, launched years ago, is traveling in a straight line away from the solar system. Which law of motion does this illustrate?

_____

Answers and explanations start on page 118.

Work

When a physicist talks about work, she is not necessarily referring to her job. In physics, **work** occurs when a force moves an object. If there is no movement, it doesn't matter how much force was applied—no work was done. So, if you push a heavy rock and it doesn't move, you have done no work. If the rock moves, then you've done work.

Energy

In the chemistry lesson, you learned that matter has mass and takes up space. But energy, such as heat or mechanical energy, does not have mass or take up space. What, then, is energy? As physics teacher Chris Webb explains on the video, **energy** is the ability to move matter, or do work.

Mechanical energy is the type of energy you saw demonstrated by the roller-coaster ride on the video. There are two types of mechanical energy.

- **Potential energy** is the stored energy of position. When a roller-coaster car is poised at the top of a hill, it has potential energy.
- **Kinetic energy** is the energy of motion. When the roller-coaster car rolls down the slope, its potential energy is changed to kinetic energy.

In a closed system like a roller coaster, the total amount of mechanical energy remains the same. It simply changes from one form to another as the car moves uphill and downhill. This is an example of the **Law of Conservation of Energy.** This law states that energy can neither be created nor destroyed, only changed in form.

Machines

Machines are devices that transmit a force, changing its size or direction. The force applied to a machine is the effort force. The force the machine overcomes is the load. The amount of work that is done with or without the machine remains the same. However, with a machine, the effort force can be less than the load it overcomes if the load is moved through a greater distance. Here are some simple machines.

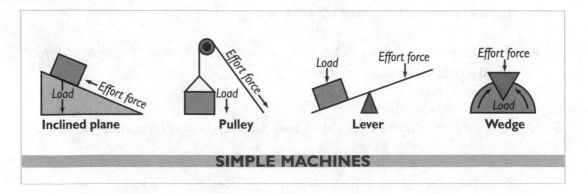

SIMPLE MACHINES

A machine helps people do work by changing the size or direction of the applied force. For example, it is very hard to lift a heavy box straight up from the ground into a truck. If you use a ramp (an inclined plane), it is much easier to push the box into the truck. Although you must do the same amount of work when lifting the box as pushing it up the ramp, the force you must exert in each situation changes. When lifting the box, you exert a large force over a small distance. When pushing the box, you exert a smaller force over a larger distance.

PHYSICS ■ PRACTICE 2

A. Use the information on pages 82 and 83 to match each term with its definition. Write the letter of the term in the space provided.

_____ 1. the ability to move matter   **a.** machine

_____ 2. the stored energy of position   **b.** work

_____ 3. the energy of motion   **c.** energy

_____ 4. a force applied to a machine   **d.** potential energy

_____ 5. a device that transmits a force,   **e.** effort force
         changing its size or direction   **f.** kinetic energy

_____ 6. the application of a force to move an object

B. Use the information on pages 82 and 83 to answer the following questions.

7. Compare the Law of Conservation of Matter (page 68) and the Law of Conservation of Energy (page 82).

_____

8. Give a real-life example of each of the machines shown on page 82.
    a. inclined plane: _____   **b.** pulley: _____
    c. lever: _____   **d.** wedge: _____

9. In the diagrams below, draw the roller-coaster car in the proper place to illustrate potential and kinetic energy.

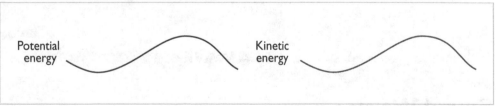

Potential energy   Kinetic energy

Answers and explanations start on page 118.

"A wave is a wiggle in space and time; like an ocean wave, it moves up and down."

Waves

If you drop a rock into a pond, ripples spread out in all directions. These ripples are **waves,** back-and-forth motions that repeat themselves. All waves are back-and-forth motions, including ocean waves, sound waves, seismic waves, and microwaves.

Waves carry energy, but they do not move matter along. For instance, a cork in the ocean will bob up and down as each wave strikes it. However, the waves will not move it far from its original position, although it could be blown around by a strong wind.

Parts of a Wave

All waves can be described using a few special terms. Look for each part in the diagram below as you read this list.

- *Crest.* The highest point of a wave is the crest.
- *Trough.* The lowest point of a wave is the trough.
- *Amplitude.* The height of the crest or depth of the trough from the midpoint is the amplitude.
- *Wavelength.* The distance between two crests or two troughs is the wavelength.

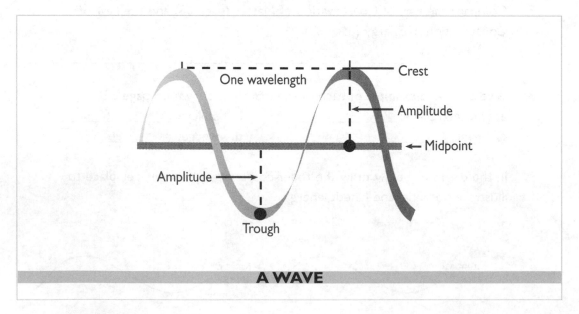

A WAVE

Sound Waves

Sound waves travel through the air, water, or ground to reach our ears. Different substances carry sound waves at different speeds. For example, it takes about 5 seconds for sound waves to travel a mile through the air. In contrast, sound waves traveling through water take a little over 1 second to go a mile. Sound waves traveling through the ground move even faster.

You can figure out how far away a lightning strike is by counting the number of seconds between the lightning and the thunder. You see the flash of lightning instantly because light waves travel almost a million times faster than sound waves. The thunder follows several seconds later, even though it occurred at the same time. Since sound waves in air travel a mile in 5 seconds, divide the number of seconds between lightning and thunder by 5. That will give you the distance in miles to the lightning strike.

PHYSICS ▪ PRACTICE 3

A. Based on the information on pages 84 and 85, write _True_ or _False_ next to each statement.

_____ **1.** Waves transfer energy but do not move matter along.

_____ **2.** The high point of a wave is its crest.

_____ **3.** The low point of a wave is its trough.

_____ **4.** The distance between one crest and the next is the amplitude.

_____ **5.** The height of the crest from the midpoint of a wave is the wavelength.

B. Use the information on pages 84 and 85 to answer the following questions.

6. Why does a beach ball or a chip of wood bob up and down on water waves?

_____

_____

7. During a thunderstorm, you see lightning. About 10 seconds later, you hear thunder. How far away is the lightning strike?

_____

8. Earthquakes produce seismic waves. Through what type of matter do seismic waves usually travel? (Refer to page 40 if you have forgotten.)

_____

Answers and explanations start on page 118.

Electromagnetic Waves

When you look around, you see objects everywhere. In reality, you are seeing light waves. The light waves come from the sun or a lamp and reflect off the objects around you.

Light waves are **electromagnetic waves.** Electromagnetic waves consist of a wave motion of electric and magnetic fields. Unlike sound waves and water waves, electromagnetic waves do not need to travel through matter. Electromagnetic waves can travel through the vacuum (complete emptiness) of space.

The Electromagnetic Spectrum

There is a whole range of electromagnetic waves called the **electromagnetic spectrum.** At one end of the spectrum are cosmic rays with very short wavelengths. At the other end of the spectrum are television and radio waves with a wavelength of 1 kilometer and longer. In the middle are light waves.

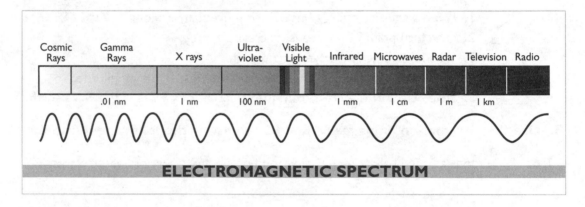

ELECTROMAGNETIC SPECTRUM

We use many portions of the electromagnetic spectrum.
- We use light waves to see, of course. We also use light waves in lamps and in laser instruments such as price scanners and those used in surgery and light shows.
- We use television and radio waves to transmit information that is interpreted as images and sounds by TVs and radio receivers.
- We use X rays to take pictures of bones. X rays work because the waves pass through the softer parts of the body but are absorbed by bone, casting a shadow on film.
- We use microwaves to cook food. Microwaves cause the water molecules in food to vibrate rapidly back and forth; this motion creates heat and warms the food.
- We use gamma rays in radiation therapy. Gamma rays destroy tumors and cancers.
- We use infrared rays in warming lights in restaurant food lines.
- We use radar waves to detect faraway unseen objects such as airplanes.

A. Use the information on page 86 to complete these sentences.

1. Light waves are a type of _____ wave.

2. An electromagnetic wave is a wave motion of _____ and _____ fields.

3. Electromagnetic waves can travel through the emptiness of _____.

4. The different types of electromagnetic waves are arranged in the _____, from waves with very short wavelengths to waves with very long wavelengths.

B. Use the information on page 86 to match each type of electromagnetic wave with one of its uses. Write the letter of the wave in the space provided. (Some answers will be used more than once.)

_____ 5. cook food by energizing water molecules

_____ 6. transmit information interpreted by receivers as sound

_____ 7. cut through tissue in laser surgery

_____ 8. scan prices at the grocery store

_____ 9. destroy cancer cells and tumors

_____ 10. detect distant, unseen objects

_____ 11. warm food in a restaurant kitchen

_____ 12. take pictures of bones

_____ 13. transmit information interpreted by receivers as pictures and sound.

a. radio waves

b. radar waves

c. microwaves

d. infrared waves

e. visible light

f. television waves

g. X rays

h. gamma rays

C. Use the information and diagram on page 86 to answer these questions.

14. What is one important difference between sound waves and electromagnetic waves?

_____

15. Approximately what is the wavelength of microwaves?

_____

Answers and explanations start on page 119.

SCIENCE SKILLS

"Electricity and magnetism are inseparable. They always go together."

Magnetism and Electricity

What Is Magnetism?

The ancient Greeks and Chinese knew that magnetite, a type of rock, attracts iron and some other metals. When an ordinary bar of iron or steel is stroked with a piece of magnetite, its atoms line up and it becomes magnetized. Once the iron or steel becomes magnetic, it behaves just like magnetite. It can be used to make more magnets.

Unmagnetized iron bar

Magnetized iron bar (bar magnet)

What is magnetism? **Magnetism** is a force produced by moving or spinning electric charges. This force is strongest at the two ends of a magnet, called the poles. If you put a bar magnet in a container of iron filings, the filings cling to each pole as shown here.

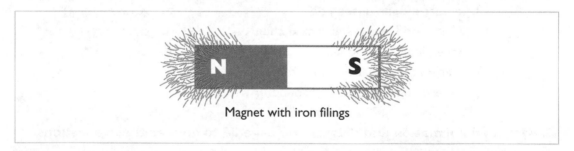
Magnet with iron filings

If you hung this magnet from a string, one end would swing to the north. This happens because Earth itself is a giant magnet with its ends near the North and South poles. The end of the bar magnet that points north is called the **north pole.** The end that points to the south is called the **south pole.** This property of magnets is used in compasses. A magnetic needle is suspended in a compass so it can swing freely and point north.

The north pole of one magnet is attracted to the south pole of another. Two south poles repel each other, and two north poles repel each other. This principle of magnetism is often stated this way: Unlike poles attract each other; like poles repel each other.

A magnet produces an area of force called a **magnetic field.** The magnetic field is strongest near the poles. However, a magnet exerts force some distance away. The stronger the magnet, the larger the area in which its magnetic field is detectable.

Uses of Magnetism

Magnetism has many uses in everyday life. Here are some examples:

- Magnetic tape is used to record sound, images, and other information. Videotape and audiotape are magnetic tapes.
- Magnetic disks also are used to record information. Computer hard drives are magnetic disks.
- Magnetic resonance imaging (MRI) gives three-dimensional images of the soft tissues of the body. MRI is used to diagnose cancer and other problems not picked up by X rays.
- Electromagnets are powerful magnets that can be turned on and off. They are used for raising heavy steel and scrap metal.

In addition, magnets are used in the generation of electricity and in electric motors.

PHYSICS ■ PRACTICE 5

A. Based on the information on pages 88 and 89, write *True* or *False* next to each statement.

_____ 1. Magnetism is a force produced by moving or spinning electric charges.

_____ 2. In a bar magnet, the atoms are arranged at random.

_____ 3. You can tell direction using a compass because the compass has a magnetic needle that is attracted to the magnetic pole of Earth.

_____ 4. A magnetic field is weakest near the poles of a bar magnet.

_____ 5. Unlike poles attract each other; like poles repel each other.

B. Use the information on pages 88 and 89 to answer these questions.

6. Describe two items you own that make use of magnets.

   _____

   _____

7. Why is it useful to be able to turn off the magnetic field of an electromagnet?

   _____

   _____

Answers and explanations start on page 119.

What Is Electricity?

All atoms have positively charged nuclei and negatively charged electrons (see pages 60 and 61). An **electric current** is the flow of electric charges, usually electrons jumping from one atom to the next.

Producing an Electric Current

One way to produce an electric current is to create a moving magnetic field, which sets electric charges in motion. In an electric generator, a coil of wire is held between the poles of a magnet. When the coil is turned in the magnetic field, electric current flows in the wire. In an electric power plant, energy is needed to turn the coil to produce current. The energy from fossil fuels, water power, or nuclear power is used to heat water to make steam. The steam drives a turbine, which turns the wire coils in the generators, producing electric current.

Another way to make an electric current is to use a battery (dry cell). A dry cell has a chemical paste and two electrodes. Electrons flow out of one electrode, through the paste, and toward the other electrode. This creates a current.

An Electric Circuit

When a battery is connected to a wire and to a device such as a light or a motor, the electric current flows from the negative electrode, through the wire and the device, and back to the positive electrode. The path along which current flows is called a **circuit.**

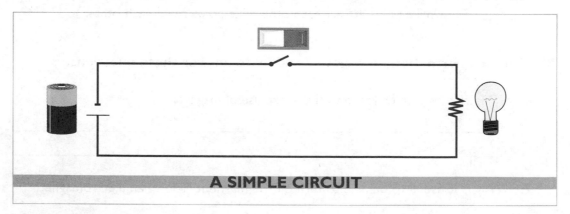

A SIMPLE CIRCUIT

The diagram above shows a simple circuit, such as the one in a flashlight. In this circuit diagram, symbols are used to represent certain types of things:

- A line represents the wire of the circuit.
- Short lines perpendicular to the wire represent the power source, or battery.
- A short hinged line represents the switch. For the flashlight to light up, the switch must be closed, so the circuit will be complete.
- Zigzag lines represent a resistor, a device that limits the flow of current. In this case, the resistor is a light bulb.

A. **Use the information on page 90 to answer the following questions. Place the letter of the correct answer in the space provided.**

_____ **1.** An electric current consists of the flow of negatively charged particles called
    a. electrons   **b.** protons
    c. neutrons   **d.** nuclei

_____ **2.** An electric generator produces electric current by turning a coil of wire in a
    a. battery   **b.** resistor
    c. turbine   **d.** magnetic field

_____ **3.** Electric current can be produced by a generator or by a
    a. battery   **b.** resistor
    c. turbine   **d.** motor

_____ **4.** An electric current flows in a path called a
    a. switch   **b.** circuit
    c. resistor   **d.** circuit diagram

_____ **5.** The purpose of a switch is to
    a. provide a source of power   **b.** slow the flow of current
    c. start and stop the current   **d.** recharge the battery

_____ **6.** In a flashlight, which part is a resistor?
    a. the battery   **b.** the case
    c. the switch   **d.** the light bulb

B. **Use the information on page 90 to answer the following questions.**

7. Is the flashlight represented by the circuit diagram on or off? Explain.

_____

_____

8. List five or more uses of electricity at home and on the job.

_____

_____

Answers and explanations begin on page 119.

Evaluate Science Information

When you **evaluate** science information, you decide whether it makes sense or not. In science, conclusions are drawn based on the evidence provided by observations, measurements, and experiments. When you read science articles, think logically. Ask yourself whether the evidence provided supports the conclusion being drawn.

On the GED Science Test, some questions will ask whether certain evidence supports a conclusion. Some will ask you to evaluate science methods and reasoning. (For more about scientific methods and thinking, see page 126 in the Science Resources section.)

When you answer evaluation questions, you decide whether conclusions are supported by particular evidence. You assess whether methods and reasoning used make sense.

EXAMPLE

The International Space Station would travel at a constant speed in a straight line away from Earth, except for the force of gravity. Gravity pulls the space station toward Earth. But the space station has a high velocity, and its momentum prevents it from falling. The interplay of the pull of gravity and the space station's momentum keeps the space station moving in a stable orbit.

In 2000, another force came into play—friction. High levels of sunspot activity heated the upper atmosphere. As the space station orbited 210 miles above Earth, friction from air molecules in the upper atmosphere increased, and the space station slowed down. It lost altitude at a rate of 2 miles per week. To correct this problem, the space shuttle *Atlantis* nudged the space station upward an extra 25 miles. This restored a stable orbit.

What evidence supports the conclusion that increased sunspot activity resulted in increased friction on the space station?
  a. The space station orbited about 210 miles above Earth.
  b. The space station started to slow down and lose altitude.

If you answered *b,* you are correct. The space station started falling during a period of high sunspot activity. It was slowing down because of friction from heated air.

THINKING STRATEGY: In this case, you must look for a fact or observation that makes the conclusion about sunspot activity a reasonable one. The fact or observation must provide evidence for the conclusion.

Now let's look at other evaluation questions similar to those you will see on the GED.

Sample GED Question

A robot is a complex machine that can move. It performs tasks by following programmed instructions in its internal computer. To change the tasks a robot performs, you have to reprogram the computer.

Robots also have sensors, which detect light, heat, shapes, motion, and sound. This information helps the robot's computer decide what to do next.

Which of the following facts supports the conclusion that a robot cannot learn new tasks as people can?

(1) A robot is able to move by following directions from its computer.

(2) A robot performs tasks according to instructions from its computer.

(3) A robot can perform new tasks only after its computer has been reprogrammed.

(4) A robot has sensors that relay data about its surroundings to its computer.

(5) The sensors of a robot are similar to the senses people have.

THINKING STRATEGY: Review the passage, looking for information that explains how a robot acquires the ability to perform new tasks. This information should support the conclusion that a robot acquires new knowledge in a way that differs from the way people acquire knowledge.

The correct answer is **(3) A robot can perform new tasks only after its computer has been reprogrammed.** This fact explains how a robot "learns" and supports the conclusion that this is different from the way people learn.

GED THINKING SKILL PRACTICE ■ EVALUATION

In people with good eyesight, the lens of the eye bends light rays so that they form an image on the retina, the rear surface of the eye. In nearsighted people, the eyeball is too long. The image forms in front of the retina, and it appears blurred.

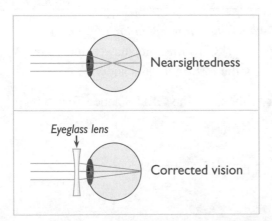

Nearsightedness

Eyeglass lens

Corrected vision

Which conclusion is supported by the information and diagram?

(1) Nearsighted people have long lenses.

(2) Light rays exit the back of the eye.

(3) Light rays enter the eye through the retina.

(4) Nearsightedness can be corrected by eyeglass lenses.

(5) Nearsightedness runs in families.

Answers and explanations start on page 120.

Understand Diagrams

On the GED Science Test, you will answer questions based on diagrams. Many science diagrams show something like the human ear or the solar system with their parts labeled (see pages 37 and 48). Others, like the circuit diagram on page 90, represent things in a more abstract way. This type of diagram shows things that are hard to draw, like forces, electricity, or light waves, in a visual way that makes them easier to understand.

EXAMPLE

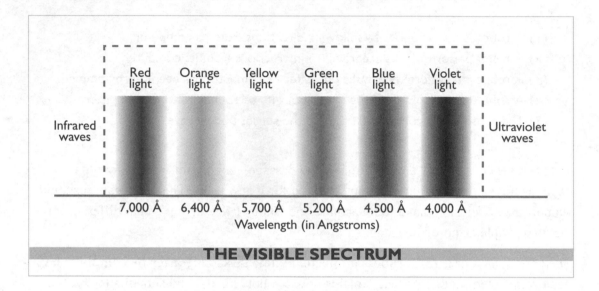

THE VISIBLE SPECTRUM

1. What is the topic of this diagram? _____

Did you answer, *the visible spectrum* or *the part of the electromagnetic spectrum we can see?* If so, you were right. The topic of a diagram is usually found in its title.

2. How does the diagram show the different wavelengths of colored light?

_____

_____

_____

The right answer is, *The diagram gives the wavelength of each color of light in angstroms and a sketch of waves of different lengths.* To find the answer, look for "Wavelength" in the diagram and examine how the information is shown.

Now let's look at diagram questions similar to those on the GED Science Test.

GED Diagram Practice

An object's velocity is its speed plus direction. An object may have more than one velocity. When a child pulls a wagon and his friend rolls a ball across the back of the wagon, the ball has two velocities. First, it has the same forward velocity as the moving wagon. Second, it has a crosswise rolling velocity. The resultant velocity, a vector, is shown by the thick arrow.

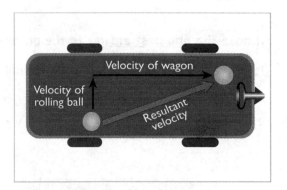

Suppose the wagon stood still while the ball was rolled across the back of the wagon. How would you change the diagram to show the ball's velocity?

(1) Cross out the wagon.

(2) Cross out the ball.

(3) Cross out the vector showing the wagon's velocity.

(4) Cross out the vectors showing the wagon's velocity and the ball's resultant velocity.

(5) Cross out the vector showing the ball's rolling velocity.

THINKING STRATEGY: First read the short passage and the question. Picture in your mind the moving wagon and the ball rolling across it. Then examine the diagram to see how this situation is shown. Take the time to figure out what the vector arrows show. Then picture the new situation described in the question. How would this be shown?

The correct answer is **(4) Cross out the vectors showing the wagon's velocity and the ball's resultant velocity.** The wagon is not moving, so its vector should be crossed out. The ball has only one velocity, so the resultant velocity can be crossed out, too.

GED GRAPHIC SKILL PRACTICE ■ USING DIAGRAMS

The following question is based on the diagram below.

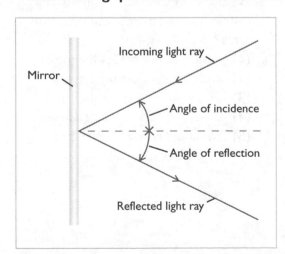

The diagram suggests that the mirror acts as

(1) a barrier through which light rays cannot pass

(2) a medium through which light rays pass

(3) an amplifier of the strength of light rays

(4) a medium in which light rays are stored

(5) a device that sorts light by color

Answers and explanations start on page 120.

GED Review: Physics

Choose the <u>one best answer</u> to the questions below.

<u>Questions 1 and 2</u> refer to the following information and diagram.

When two surfaces move past one another, their roughness causes friction. When separated by a lubricant, the surfaces slide more easily past one another.

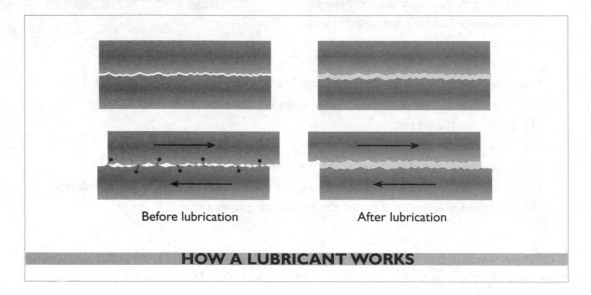

Before lubrication   After lubrication

HOW A LUBRICANT WORKS

1. A lubricant works by
   - **(1)** decreasing gravity
   - **(2)** decreasing the surface's roughness
   - **(3)** increasing the surface's roughness
   - **(4)** reducing friction
   - **(5)** increasing friction

2. Lubricants are most useful for
   - **(1)** preventing wear on machine parts
   - **(2)** preventing objects from slipping
   - **(3)** making abrasives like sandpaper
   - **(4)** wearing away rough surfaces
   - **(5)** separating sheets of paper

3. When an object moves in a circle, the force acting on it is called centripetal force. Without this force, the object would fly off in a straight line. For example, mud flies off a spinning tire, moving in a straight line away from the tire.

   A teacher wants to demonstrate centripetal force for his class. He ties a string to a cork and drags the cork in a straight line along the floor.

   What should the teacher have done instead to demonstrate centripetal force?
   - **(1)** push the cork across the table
   - **(2)** push the cork off the table
   - **(3)** throw the cork up in the air
   - **(4)** swing the cork like a pendulum
   - **(5)** swing the cork in circles and then let go

Questions 4 and 5 refer to the following table.

Sound	Decibels
Whispering	10–25
Automobile	40–50
Conversation	50–60
Heavy street traffic	70–90
Riveting gun	90–100
Thunder	110
Jet aircraft	130
Rocket takeoff	150

☐ Loudness causes pain.

LOUDNESS OF COMMON SOUNDS (IN DECIBELS)

4. Which of the following sounds has a decibel level of about 50 to 60?
   (1) whispering
   (2) an automobile
   (3) conversation
   (4) heavy street traffic
   (5) a riveting gun

5. People who are regularly exposed to sounds of more than 80 decibels for a long time without ear protection are at risk of damaging their hearing.

   Which of the following people would be at the highest risk for hearing damage if they didn't wear ear protection?
   (1) cooks in a restaurant
   (2) office workers
   (3) truck drivers
   (4) airport ground workers
   (5) meteorologists

Questions 6 and 7 refer to the following information.

When we see the color of an object, we really perceive the particular color light waves that it reflects. For example, an object that looks green absorbs all the color light waves except those of green light. An object that looks white reflects almost all of the light waves that hit it. That's because white light is a blend of light waves of all the colors in the visible spectrum.

Sunlight and fluorescent light are both white light. However, they are slightly different combinations of light waves of the colors.

6. Why do different objects appear to be different colors?
   (1) They are made of different substances.
   (2) They reflect different color light waves.
   (3) They are actually white.
   (4) They are actually black.
   (5) The primary colors can be combined.

7. Which of the following is evidence to support the statement that sunlight and fluorescent light have different combinations of color light waves although they both appear to us as white light?
   (1) Fluorescent light is artificial and sunlight is natural.
   (2) Incandescent light looks more like sunlight than fluorescent light does.
   (3) Clothing appears to be different colors in fluorescent light and sunlight.
   (4) Under fluorescent light, objects absorb all the color light waves.
   (5) Sunlight appears to be different colors at different times of day.

Answers and explanations begin on page 120.

GED REVIEW

Science Posttest

The Science Posttest on the following pages is similar to the GED Science Test. However, it has only 25 items, compared to 50 items on the actual GED Science Test.

This Posttest consists of short passages, charts, tables, diagrams, and graphs about science. Each passage or graphic is followed by one or more multiple-choice questions. Read each passage, study the graphics, and then answer the questions. You may refer back to the passage or graphic at any time.

The purpose of the Posttest is to evaluate your science knowledge and thinking skills. The Posttest will help you identify the content areas and skills that you need to review.

Directions

1. Read the sample passage and test item on page 99 to become familiar with the test format.

2. Take the test on pages 100 through 107. Read each passage, study the graphics, if any, and then choose the best answer to each question.

3. Record your answers on the answer sheet below, using a No. 2 pencil.

4. Check your work against the Answers and Explanations on page 108.

5. Enter your scores in the evaluation charts on page 109.

SCIENCE POSTTEST ▪ ANSWER SHEET

Name _____   Date _____

Class _____

1. ①②③④⑤	6. ①②③④⑤	11. ①②③④⑤	16. ①②③④⑤	21. ①②③④⑤
2. ①②③④⑤	7. ①②③④⑤	12. ①②③④⑤	17. ①②③④⑤	22. ①②③④⑤
3. ①②③④⑤	8. ①②③④⑤	13. ①②③④⑤	18. ①②③④⑤	23. ①②③④⑤
4. ①②③④⑤	9. ①②③④⑤	14. ①②③④⑤	19. ①②③④⑤	24. ①②③④⑤
5. ①②③④⑤	10. ①②③④⑤	15. ①②③④⑤	20. ①②③④⑤	25. ①②③④⑤

Sample Passage and Test Item

The following passage and test item are similar to those you will find on the Science Posttest. Read the passage and the test item. Then go over the answer sheet sample and explanation of why the correct answer is correct.

Question 0 refers to the following table.

Type of Cell	Lifespan
Red blood cells	120 days
Bone cells	25–30 years
Brain cells	Lifetime
Colon cells	3–4 days
Skin cells	19–34 days
Sperm cells	2–3 days

LIFESPAN OF HUMAN CELLS

0. According to the table, which type of cell has the shortest lifespan?
   (1) red blood cells
   (2) bone cells
   (3) colon cells
   (4) skin cells
   (5) sperm cells

Marking the Answer Sheet

0. ①②③④⑤

The correct answer is **(5) sperm cells.** Therefore, answer space 5 is marked on the answer sheet, as shown above. The space should be filled in completely using a No. 2 pencil. If you change your mind about an answer, erase it completely.

Answer and Explanation

(5) sperm cells (Comprehension) In the column labeled "Lifespan," look for the shortest amount of time. That is 2–3 days, the length of time that sperm cells live.

Science Posttest

Choose the <u>one best answer</u> to the questions below.

<u>Questions 1 and 2</u> refer to the following passage.

After giving birth, many mothers make a medical donation of the blood from the umbilical cord and placenta. This cord blood is very valuable because it contains blood stem cells. They can develop into other types of blood cells—red blood cells that carry oxygen, white blood cells that fight infection, and platelets that clot blood.

The stem cells in cord blood can treat several diseases. They can rebuild the blood of a child with leukemia. They can help a patient with sickle cell anemia. And they can help restore the white blood cells of babies with immune system problems.

1. What is a blood stem cell?
(1) a type of white blood cell
(2) a type of red blood cell
(3) a type of platelet
(4) a cell that can develop into other types of blood cells
(5) a cell from a new mother

2. Which of the following conclusions is supported by the information in the passage?
(1) Companies should store cord blood.
(2) Cord blood can help treat patients with many different blood-related diseases.
(3) Cord blood is better than adult blood.
(4) Most mothers are opposed to donating cord blood for the use of others.
(5) Babies need their cord blood to survive.

3. Habitats can be destroyed either by natural causes such as climate change or by human activity such as development.

Which of the following is an example of habitat destruction by natural means?
(1) a river dammed to generate power
(2) a meadow plowed for farming
(3) woods cleared for building houses
(4) a barrier island washed away by a hurricane
(5) a river polluted with agricultural runoff

4. Children with high exposure to lead paint can develop learning problems. High lead levels are sometimes treated with a drug called succimer. Succimer flushes lead out of the body, but it can have serious side effects. Scientists studied 780 children with high lead levels. One group took succimer. The other group took a dummy medication. The lead levels in the children taking succimer dropped much more than those in the children in the other group. However, after three years, there was no difference in the I.Q. levels of the two groups.

Which data provides evidence that succimer is not effective in preventing problems caused by high levels of lead?
(1) Succimer flushes lead out of the body.
(2) Succimer can cause side effects.
(3) Children who took a dummy medication still had high levels of lead.
(4) Children who took succimer had decreased lead levels.
(5) Children in both groups had similar I.Q. levels after three years.

Traditional vaccines must be refrigerated, and they last only about six months. In contrast, experimental DNA vaccines can be stored at room temperature and last indefinitely. The diagram shows how each type of vaccine is administered.

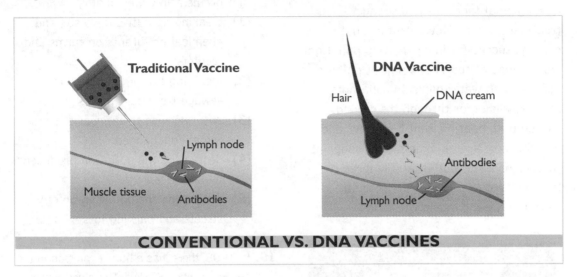

CONVENTIONAL VS. DNA VACCINES

5. Which statement best summarizes the main idea of the information and diagram?
   (1) DNA vaccines can be stored at room temperature.
   (2) Scientists are experimenting with DNA vaccines.
   (3) Experimental DNA vaccines have several advantages over traditional vaccines.
   (4) Traditional vaccines are injected and DNA vaccines are applied to the skin.
   (5) Traditional vaccines last only a few months.

6. Suppose that DNA vaccines are approved for use with humans. From a patient's point of view, what would be the main advantage of the DNA vaccine?
   (1) It can be stored at room temperature.
   (2) It lasts indefinitely.
   (3) It contains DNA rather than proteins.
   (4) It is applied to the skin rather than injected.
   (5) It is experimental.

7. The DNA vaccine seeps into the skin along the hair. To which of the following is this most similar?
   (1) a soft contact lens placed on the eye to correct vision
   (2) a padded cushion placed on a corn or a callous to relieve pain
   (3) a strip bandage placed on a cut to keep it clean
   (4) a butterfly bandage placed across a cut to pull the two sides together for healing
   (5) a patch placed on the skin to provide smokers with small amounts of nicotine

Questions 8 and 9 refer to the following information and table.

The United States has cleaned up much water pollution from point sources, such as municipal factories and sewage treatment plants. However, it has not been as successful in cleaning up pollution from nonpoint sources. Nonpoint pollution occurs when water picks up pollutants as it travels on or through the ground. Agriculture, boating, septic systems, and urban runoff are all examples of nonpoint sources of water pollution. The three leading sources of water pollution are shown in the table below.

Rank	River Pollution	Lake Pollution	Estuary* Pollution
1	Agriculture	Agriculture	Urban runoff
2	Municipal point sources	Municipal point sources	Municipal point sources
3	Stream/habitat changes	Urban runoff	Agriculture

*A coastal area where fresh and sea water mix.

LEADING SOURCES OF WATER POLLUTION

Source: U.S. Environmental Protection Agency

8. What is the main source of water pollution in estuaries?
   (1) agriculture
   (2) urban runoff
   (3) municipal point sources
   (4) stream/habitat changes
   (5) factories

9. About 40% of U.S. waterways are so polluted that fishing and swimming are not safe. Which of the following actions would be likely to cause the greatest improvement in water quality?
   (1) cleaning up nutrient, waste, and chemical pollutants on farms and feedlots
   (2) installing filtering systems in sewage treatment plants
   (3) ensuring that streams and rivers remain unchanged
   (4) decreasing the pollution produced by factories
   (5) decreasing the pollution on city streets and parking lots

10. In 1908, there was a huge explosion in the air over central Siberia. The blast was as powerful as a hydrogen bomb. It leveled thousands of square miles of forest.

   Five theories were proposed to explain this event: (1) A large meteorite hit Earth; however, there was no crater at the site of the blast. (2) A piece of anti-matter hit Earth, but there was no radiation, the usual result of the collision of matter and anti-matter. (3) A small black hole passed through Earth; however, there was no corresponding explosion on the other side of the globe. (4) An alien spacecraft crashed into Earth, but no debris from such a craft was ever found. (5) A comet entered the atmosphere, creating a fireball and shock wave; the comet melted before reaching the ground.

   Based on the information provided, which theory has the best supporting evidence?
   (1) a large meteorite
   (2) a piece of anti-matter
   (3) a small black hole
   (4) an alien spacecraft
   (5) a comet

Questions 11 and 12 refer to the following graph.

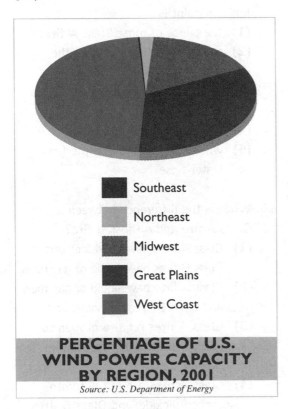

PERCENTAGE OF U.S. WIND POWER CAPACITY BY REGION, 2001

Source: U.S. Department of Energy

Legend:
- Southeast
- Northeast
- Midwest
- Great Plains
- West Coast

11. Which region had about one-third of the U.S. wind power capacity in 2001?
(1) the West Coast
(2) the Southeast
(3) the Northeast
(4) the Midwest
(5) the Great Plains

12. Peter and Mariah are starting a business that services windmills and other equipment used to generate electricity from wind power.

In which region are they likely to find the most work?
(1) the West Coast
(2) the Southeast
(3) the Northeast
(4) the Midwest
(5) the Great Plains

Question 13 refers to the following information.

The Saffir-Simpson Hurricane Damage-Potential scale measures the disaster potential of hurricanes. The scale includes air pressure, wind speed, and water surge measurements. It also describes damage in these categories:

Minimal No structural damage. Some tree, shrubbery, and mobile home damage.

Moderate Some roof, window, and door damage. Considerable damage to vegetation, mobile homes, and piers.

Extensive Some structural damage to small or residential buildings. Mobile homes destroyed.

Extreme Extensive roof, window, and door damage. Major damage to lower floors of structures. Some roof failure on small residences.

Catastrophic Complete roof failure in many buildings. Some complete building failures. Major damage to lower floors of all structures within 500 yards of shore.

13. In Hurricane Lester, Beatrice's mobile home was completely demolished. Her friend Duane's small house suffered some structural damage but withstood the storm. In which damage category would you place Hurricane Lester?
(1) minimal damage
(2) moderate damage
(3) extensive damage
(4) extreme damage
(5) catastrophic damage

14. Density is the ratio of the mass of an object to its volume. Sheriah did a project to show that different substances have different densities. She assembled five substances: a brick, a piece of plastic foam, a feather, a one-pound weight, and a rubber ball. She arranged the items in order of weight and made a sign that said, "Five Substances in Order of Increasing Density."

When Sheriah's mother looked at the project, she suggested that Sheriah should have done something differently to make the display correct. What was it?

(1) Sheriah should have used five samples of the same substance.

(2) Sheriah should have selected five liquids rather than five solids.

(3) Sheriah should have used samples of equal volume.

(4) Sheriah should have used samples of the same shape.

(5) Sheriah should have used a solid, a liquid, and a gas.

Questions 15 and 16 refer to the following passage.

Fire is the rapid reaction of a fuel with oxygen. Most fires involve the combustion of a solid fuel—for example, a burning wood-framed house. This type of fire is known as a Class A fire. Firefighters use water to put out Class A fires. Water cools the fuel to a temperature below its ignition point. It also turns to steam; steam dilutes the oxygen in the air. With less oxygen, the combustion reaction slows down.

The fuel of a Class B fire is a liquid or a gas. Water is not used to extinguish Class B fires because water can spread the fuel around. Carbon dioxide or dry chemical fire extinguishers are used to put out Class B fires.

15. From the information in the passage, you can infer that a substance's ignition point is

(1) the cause of most Class A fires

(2) the temperature at which the substance begins to burn

(3) the temperature at which flames start to spread

(4) where the fire actually starts

(5) where firefighters aim the first water hose

16. What is the difference between a Class A fire and a Class B fire?

(1) Class A fires have solid fuel, and Class B fires have liquid or gas fuels.

(2) Class A fires have liquid or gas fuels, and Class B fires have solid fuel.

(3) Class A fires require oxygen to burn, and Class B fires require carbon dioxide.

(4) Class A fires are put out with carbon dioxide, and Class B fires are put out with water.

(5) Class A fires are spread by water, and Class B fires are spread by wind.

Questions 17 through 19 refer to the following information and table.

Some starchy foods cooked at high temperatures contain the chemical acrylamide, as shown in the table below. Until recently, this substance was known as an industrial chemical that may cause cancer. Chemists do not know how acrylamide forms during cooking. From the few data there are, it seems that the higher the temperature, the more acrylamide forms.

The Environmental Protection Agency sets a limit of 0.12 micrograms of acrylamide per 8-ounce glass of water. The amount in foods is not regulated.

Food	Acrylamide per Serving (micrograms)
Oat cereal, 1 oz.	7
Tortilla chips, 1 oz.	5
Taco shells (3), 1.1 oz.	1
Boiled potatoes, 4 oz.	0
French fries (baked), 3 oz.	28
Potato chips, 1 oz.	25
Fast-food french fries, 5.7 oz.	57

ACRYLAMIDE IN STARCHY FOODS

17. How many micrograms of acrylamide were found in a serving of tortilla chips?
    (1) 0
    (2) 1
    (3) 5
    (4) 7
    (5) 28

18. Which foods have more acrylamide than is allowed in an 8-ounce glass of water?
    (1) all of them
    (2) all of them except boiled potatoes
    (3) baked french fries only
    (4) fast-food french fries only
    (5) tortilla chips and taco shells only

19. After reading a news article about acrylamide, Karen decided to make her family's diet safer. Most likely, Karen cut back on
    (1) french fries
    (2) oat cereal
    (3) tortilla chips
    (4) taco shells
    (5) boiled potatoes

Question 20 refers to the following information and diagram.

The sound of a gently vibrating tuning fork would have a wave pattern like this:

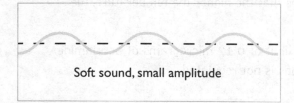

Soft sound, small amplitude

The sound of the same tuning fork vibrating more strongly would have a wave pattern like this:

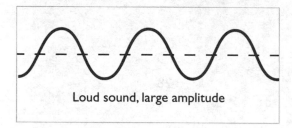

Loud sound, large amplitude

20. A tuning fork always gives off a sound at the same pitch. What is the difference between the sound of the gently vibrating and the strongly vibrating tuning fork?
   (1) The strongly vibrating tuning fork makes a louder sound.
   (2) The strongly vibrating tuning fork makes a softer sound.
   (3) The strongly vibrating tuning fork makes a higher-pitched sound.
   (4) The strongly vibrating tuning fork makes a lower-pitched sound.
   (5) Only the strongly vibrating tuning fork makes any sound.

21. The usual way scientists identify gases is by collecting samples on site. Recently, using a specific instrument—a special type of infrared spectrometer, scientists were able to identify gases without gathering samples on site. From miles away, the spectrometer can measure the sun's infrared waves as they are absorbed by the gases. Different gases absorb different frequencies of infrared waves. So their composition can be determined from afar.

Which of the following would be a good use of this special kind of infrared spectrometer?
   (1) to identify the gases present in a lab
   (2) to identify the gases produced by an erupting volcano
   (3) to detect natural gas deep in an underground well
   (4) to detect radioactive substances beneath Earth's surface
   (5) to determine the bacterial content of exhaled air

22. Newton's third law of motion is sometimes stated, "For every action, there is an equal and opposite reaction." This means that if one object exerts a force on another object, the second object will exert an equal and opposite force on the first.

Which of these is an example of this law?
   (1) An SUV needs more force to accelerate to 60 miles an hour than a subcompact does.
   (2) Friction and gravity slow a bullet so it eventually comes to rest.
   (3) A ship's propeller moves the ship forward as it throws water backward.
   (4) A spacecraft passes Jupiter and continues its journey out of the solar system.
   (5) A golf ball rests on a tee until someone hits it.

Questions 23 through 25 refer to the following information and graph.

Heat is the energy created by the random motion of molecules. Heat moves from warmer to colder areas. One way it moves is through conduction, by transferring energy from one area to another. Some substances, like metals, are good conductors. Heat moves rapidly through them. Others, like cloth and air, are insulators. These substances are resistant to heat flow.

In the building industry, insulators are given an R-value that indicates their resistance to heat flow. The greater the R-value, the higher the resistance.

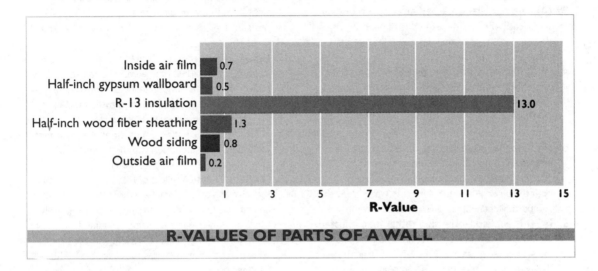

R-VALUES OF PARTS OF A WALL

23. The main purpose of insulating a building is to
  (1) provide greater structural support
  (2) ensure safe electrical wiring
  (3) improve the appearance of the exterior
  (4) improve the heat flow through walls
  (5) cut heating and air-conditioning costs

24. According to the graph, which substance provides the most resistance to heat flow?
  (1) inside air film
  (2) half-inch gypsum wallboard
  (3) R-13 insulation
  (4) half-inch wood fiber sheathing
  (5) wood siding

25. Which of the following conclusions is supported by the data in the table?
  (1) The wall contains 13 inches of insulation.
  (2) The wall is an exterior wall of a building.
  (3) The wall is an interior load-bearing wall.
  (4) Half-inch gypsum wallboard is a better insulator than half-inch wood fiber sheathing.
  (5) Metal is a good conductor of heat.

Answers and explanations begin on page 108.

Science Posttest Answers and Explanations

1. **(4) a cell that can develop into other types of blood cells** (Comprehension) According to the first paragraph, blood stem cells found in cord blood can develop into red blood cells, white blood cells, and platelets.

2. **(2) Cord blood can help treat patients with many different blood-related diseases.** (Evaluation) The passage contains information about the use of cord blood stem cells in treating leukemia, sickle-cell anemia, and infant immune problems.

3. **(4) a barrier island washed away by a hurricane** (Application) Of all the examples of habitat destruction, only this one has natural rather than human causes.

4. **(5) Children in both groups had similar I.Q. levels after three years.** (Evaluation) Although succimer does lower lead levels, it doesn't appear to enable the body to repair brain damage already done by high lead levels.

5. **(3) Experimental DNA vaccines have several advantages over traditional vaccines.** (Comprehension) This statement summarizes both the passage and the diagrams, which show the comparative benefits of DNA vaccines.

6. **(4) It is applied to the skin rather than injected.** (Analysis) Patients would not be concerned about the storage advantages of the DNA vaccine. Instead, they would be interested in a "no pain" vaccination.

7. **(5) a patch placed on the skin to provide smokers with small amounts of nicotine** (Application) Of all the choices, this is the only one that involves a substance seeping into the skin, as the DNA vaccine does.

8. **(2) urban runoff** (Comprehension) Look for the "Estuary Pollution" column. Then find the first-ranked source of water pollution in estuaries: urban runoff.

9. **(1) cleaning up nutrient, waste, and chemical pollutants on farms and feedlots** (Analysis) Since agriculture is the top source of pollutants in rivers and lakes, reducing the pollution on farms and feedlots would improve water quality the most.

10. **(5) a comet** (Evaluation) Only the comet theory is supported by all the facts—a large explosion but no physical remains of the object that caused it.

11. **(5) the Great Plains** (Comprehension) You can judge by eye which portion of the circle graph is about one-third of the whole. That is the segment representing the Great Plains.

12. **(1) the West Coast** (Application) A business that services wind power equipment would do best in the region with the most capacity—the West Coast. The West Coast has almost half the nation's capacity for wind power.

13. **(3) extensive damage** (Application) The description for "extensive damage" includes damage to small buildings (Duane's house) and destruction of model homes (Beatrice's mobile home).

14. **(3) Sheriah should have used samples of equal volume.** (Evaluation) To compare densities, you need to weigh samples that have the same volume. Sheriah had simply weighed the samples without regard to their volume.

15. **(2) the temperature at which the substance begins to burn** (Comprehension) From the information in the first paragraph, you can infer that this is the meaning of "ignition point."

16. **(1) Class A fires have solid fuel, and Class B fires have liquid or gas fuels.** (Analysis) The type of fuel is the basis for the division into Class A and Class B fires.

17. **(3) 5** (Comprehension) Locate tortilla chips in the "Food" column. Then read across the row to see how many micrograms of acrylamide the chips contain.

18. **(2) all of them except boiled potatoes** (Analysis) The passage indicates that an 8-ounce glass of water has a limit of 0.12 micrograms. Of all the foods in the table, only boiled potatoes, with no acrylamide, contain less than the limit for water.

19. **(1) french fries** (Application) Of all of the foods in the table, the french fries (whether they are baked or bought at a fast-food restaurant and so, most likely fried) have the highest levels of acrylamide per serving. Therefore, it makes sense to cut back on both types of french fries to reduce the health risk they pose.

20. **(1) The strongly vibrating tuning fork makes a louder sound.** (Analysis) According to the diagrams, the strongly vibrating tuning fork has a larger amplitude, which means it produces a louder sound.

21. **(2) to identify the gases produced by an erupting volcano** (Application) Note that this spectrometer works by measuring the absorption of the sun's infrared rays by a gas. So it cannot be used to measure underground gases or gases located any other place where sunlight doesn't reach. When a volcano begins to erupt, it gives off gases into the air. Gathering these gases is extremely dangerous. The spectrometer would enable scientists to identify gases from a distance, without risk.

22. **(3) A ship's propeller moves the ship forward as it throws water backward.** (Application) This is an example of Newton's third law of motion. The other choices are examples of the first and second laws of motion.

23. **(5) cut heating and air-conditioning costs** (Analysis) In the winter, heat loss means more energy for heating is needed. In the summer, heat gain means more energy for cooling is needed. Both cost more money.

24. **(3) R-13 insulation** (Comprehension) The substance with the most resistance to heat loss is the one with the highest R-value. The graph shows that this is R-13 insulation.

25. **(2) The wall is an exterior wall of a building.** (Evaluation) You can tell that this is the case because the wall includes wood siding and an outside air film.

Evaluation Charts for Science Posttest

Follow these steps for the most effective use of this chart:

- Check your answers against the Answers and Explanations on page 108.
- Use the following charts to circle the questions you answered correctly.
- Total your correct answers in each row (across) for science subject areas and each column (down) for thinking skills.

You can use the results to determine which subjects and graphics skills you need to focus on.

- The column on the left of the table indicates the KET Pre-GED video program and its corresponding lesson in this workbook.
- The column headings—*Comprehension, Application, Analysis,* and *Evaluation*—refer to the type of thinking skills needed to answer the questions.

SUBJECT AREAS AND THINKING SKILLS

Program	Comprehension (pp. 32–33)	Application (pp. 52–53)	Analysis (pp. 72–73)	Evaluation (pp. 92–93)	Total for Science Subjects
16 Life Science (pp. 18–37)	1, 5	3, 7	6	2, 4	___/7
17 Earth and Space Science (pp. 38–57)	8, 11	12, 13	9	10	___/6
18 Chemistry (pp. 58–77)	15, 17	19	16, 18	14	___/6
19 Physics (pp. 78–97)	24	21, 22	20, 23	25	___/6
Total for Skills	___/7	___/7	___/6	___/5	

Many of the questions on the GED Science Test are based on charts, tables, diagrams, and graphs.

- Use the chart below to circle the graphics-based questions that you answered correctly.
- Identify your strengths and weaknesses in interpreting graphics by counting the number of questions you got correct for each type of graphic.

GRAPHIC SKILLS

Diagrams (pp. 34–35, 94–95)	Charts and Tables (pp. 74–75)	Graphs (pp. 54–55)	Total for Graphics
5, 6, 7, 20	8, 9, 17, 18, 19	11, 12, 24, 25	___/13

Answers and Explanations

PROGRAM 16
LIFE SCIENCE

Practice 1 (page 21)

1. **True**

2. **False**   The human body has many different types of cells, such as skin cells and blood cells.

3. **True**

4. **True**

5. **True**

6. **False**   As you can see in the diagram of a plant cell, the chloroplasts are contained in the cytoplasm.

7. **True**

8. **True**

Your answers should be similar to the ones below.

9. Plant cells have cell walls, and animal cells do not.

10. Plant cells have chloroplasts, and animal cells do not.

Practice 2 (page 23)

1. **Water**   This comes from the soil, through the plant's roots.

2. **Carbon dioxide**   This comes from the air.

3. **Sugars**   This is the food that plants produce during photosynthesis.

4. **In all cells**   Respiration is the process that releases energy for cells to use.

5. **Sugars**   This comes from the food the organism eats.

6. **Oxygen**   This comes from the air (or from the water, in the case of fish).

7. **Carbon dioxide**   This waste product of cellular respiration is released into the air.

8. **Water vapor**   This waste product of cellular respiration is released into the air.

9. The green plant would eventually die in a dark place because it would not get any light, which is necessary for photosynthesis—the process by which the plant makes the food it needs.

Practice 3 (page 25)

1. **b. the peripheral nervous system**
   The nerves in the peripheral nervous system connect all parts of the body to the central nervous system.

2. **c. the central nervous system**   The central nervous system, consisting of the brain and the spinal cord, serves as the body's control center.

3. **c. breathing**   The brain stem controls vital functions such as breathing and heartbeat.

4. **(1) cell, (4) system, (2) tissue, (3) organ**
   Cells form tissues, tissues form organs, and organs form body systems.

Practice 4 (page 27)

1. **oxygen, carbon dioxide**   The purpose of the respiratory system is to enable the body to exchange carbon dioxide for oxygen, using the lungs.

2. **nasal cavity, trachea**   The nasal cavity refers to the nose, and the trachea refers to the windpipe; both of these structures are part of the respiratory system, and they channel air to the lungs.

3. **alveoli**   These are the tiny air sacs in the lungs where gas exchange takes place.

4. **oxygen, nutrients; carbon dioxide**   The transport of substances throughout the body is the main function of the circulatory system.

5. **heart**

6. **Arteries, veins**   The diagram shows arteries in red and veins in blue.

7. **plasma**   This is the liquid part of the blood.

8. The blood (circulatory system) absorbs oxygen from the lungs (respiratory system) and transports it throughout the body. The blood then brings waste carbon dioxide to the lungs, where it is released into the air.

Practice 5 (page 29)

1. **f. consumer**

2. **b. environment**

3. **d. ecosystem**

4. **c. producer**

5. **a. predator**

6. **e. biodiversity**

7. If the number of frogs decreased, then the number of primary consumers (insects) would increase, because they are not being eaten by frogs. In addition, the number of snakes might decrease, because one of their sources of food has decreased.

8. The consumers would have to move into another ecosystem or die from lack of food.

Practice 6 (page 31)

1. **habitat**

2. **farming, urbanization, logging**

3. **enemy** or **predator**   Without an enemy to control its numbers, the population of the nonnative species would increase tremendously.

4. **extinct**

5. **78**   Find the "Endangered Species" column and read down until you find the row for birds. The number in that box is 78.

6. **flowering plants**   Find the "Threatened Species" column and read down until you find the number 144. Then read across the row to the first column. That tells you that there are 144 threatened species of flowering plants.

7. Builders and wildlife preservationists often conflict because builders destroy habitat when they build, and wildlife preservationists want habitat to be saved for wildlife.

GED Thinking Skill Practice: Comprehension (page 33)

1. **(4) near people**   The passage indicates that the monkeys are moving to farm areas and towns. Therefore, they are moving near people.

2. **(3) food was easy to get**   According to the passage, monkeys eat crops and garbage. These types of food were more plentiful than the food they had to find in the wilderness. With a good source of food, the monkey population increased.

GED Graphic Skill Practice: Using Diagrams (page 35)

1. **(2) three** (Comprehension)   The diagram shows that a sperm cell has three main sections: the head, the middle section, and the tail.

2. **(5) flicking its tail back and forth** (Comprehension)   According to the paragraph, sperm cells whip their tails back and forth in order to move.

GED Review (pages 36–37)

1. **(2) is an ongoing process that never stops** (Comprehension)   The word *cycle* indicates a repeating process—one that continues without stopping.

2. **(5) photosynthesis** (Comprehension)   Study the diagram. Note that the one arrow pointing away from the carbon dioxide in the atmosphere shows how carbon moves back into living things; the arrow points to plant photosynthesis, the process in which plants use carbon dioxide to make food.

3. **(3) a protein** (Comprehension)   CaMK is defined in the second sentence of the paragraph as a protein.

4. **(1) The mice became stronger without exercise.** (Comprehension)   According to the passage, the extra CaMK made the mice's muscles stronger without the mice having to exercise.

5. **(5) the outer ear** (Comprehension)   According to the passage, the outer ear collects sounds. Its shape is like a funnel, so it channels sounds down the auditory canal to the eardrum.

6. **(2) the eardrum** (Comprehension)   If you study the diagram, you will see that a doctor checking an ear can look down the auditory canal as far as the eardrum, where the canal ends.

7. **(3) close to 60 percent** (Comprehension)   According to the paragraph, a report indicated that 58 percent of coral reefs are threatened by human activity.

PROGRAM 17
EARTH AND SPACE SCIENCE

Practice 1 (page 41)

1. seismic

2. fault

3. **west**   Find North America on the map, and then use the map's compass rose to identify along which coast most volcanoes and most faults are located. This is the west coast.

4. volcano

5. crust, plates

6. boundaries

7. The word *ring* refers to the arc formed by the volcano/earthquake zone surrounding the Pacific. *Ring of Fire* is a good name because, during eruptions, volcanoes spill or forcefully eject hot, fiery magma.

8. According to the map, most of the plate boundaries are located in the oceans.

Practice 2 (page 43)

1. **False**   Meteorologists cannot accurately forecast the weather a month ahead; they can make accurate forecasts only a few days ahead.

2. **False**   Rain and snow are forms of precipitation.

3. **True**

4. **False**   A maritime tropical air mass is warm and humid.

5. **True**

6. **July**   To answer the question, look for the highest point of the graph's trend line. Then see which month on the horizontal axis this high temperature corresponds to.

7. Louisville has a temperate climate. The graph shows that the mean temperature is low during the winter and high during the summer. A tropical climate, in contrast, has high temperatures throughout the year.

Practice 3 (page 45)

Nonrenewable (listed in any order)

1. **coal**

2. **natural gas**

Renewable (listed in any order)

3. **solar power**

4. **water power**

5. **wind power**

6. **geothermal energy**

   alternative answer: wood

7. **natural gas**   Note that the largest wedge on the graph represents homes heated by natural gas, at 53% of the total.

8. Wood is renewable because more trees can be planted to replace those used for firewood.

9. Most renewable sources of energy are used to generate electricity.

10. The use of renewable resources is likely to increase because fossil fuels will become scarcer.

Practice 4 (page 47)

1. **a. new air pollutant standards**   The Clean Air Act set new standards for the amounts of various air pollutants in the air.

2. **c. the air quality index**   This system alerts people when air pollution levels pose a health risk.

3. **b. Code Yellow**   Find the column that lists the air quality index numbers, looking for the row that includes the number 75. This is Code Yellow.

4. **d. Code Red**   Look at the column "Recommended Actions," and find the recommendation that everyone should limit outdoor activity. Then read across the row to see which code level the recommendation corresponds to; it is Code Red.

5. **c. summer**

6. Heavy exertion increases your breathing rate because muscles need more oxygen to produce energy through cellular respiration. When you breathe more heavily, you take in more air and also more pollution. That's why heavy exertion should be limited on days when the air is polluted.

Practice 5 (page 49)

1. f. Saturn

2. i. Pluto

3. a. Mercury

4. b. Venus

5. e. Jupiter

6. d. Mars

7. c. Earth

8. h. Neptune

9. g. Uranus

10. Pluto If you examine the diagram, you will see that Pluto orbits beyond Neptune's orbit for part of its year and inside Neptune's orbit for the other part of its year.

Practice 6 (page 51)

1. True

2. False Scientists are searching for other stars that might have planets similar to Earth; such planets, scientists infer, are most likely to support life.

3. True

4. False Stars like the sun are made mostly of hydrogen and helium. Planets and life forms have more iron, oxygen, and carbon than hydrogen and helium.

5. False Planets that support life are most likely to exist around stars midway out from the center of the galaxy, where other stars and dust are not as dense as they are in the galaxy's center.

6. True

7. True

8. Spiral and barred spiral galaxies are both shaped like disks with spiral arms. A barred spiral galaxy has a bright bar of dense stars through its center, and the spiral arms come off the ends of the bar. In contrast, a spiral galaxy has a bulging nucleus, and the spiral arms come off of the nucleus.

9. The Milky Way is a spiral galaxy.

GED Thinking Skill Practice:
Application (page 53)

(5) sandblasting a building to clean it
This is the only choice involving blown sand.

GED Graphic Skill Practice:
Using Graphs (page 55)

1. (5) 15 (Comprehension) A pictograph uses symbols or pictures to show amounts. Look at the graph's key. It tells you that one erupting volcano picture equals 5 eruptions. Since the row for Trident has 3 volcano pictures, Trident has erupted 3 × 5 times, or 15 times.

2. (3) Redoubt (Comprehension)
The key tells you that one volcano picture equals 5 eruptions. Therefore, 7 eruptions would be shown by one whole volcano (representing 5 eruptions) and one partial volcano (representing 2 eruptions). The only volcano that has one whole and one partial volcano picture in its row is Redoubt.

GED Review (pages 56–57)

1. **(4) Alabama** (Comprehension) The number of hurricanes appears on the vertical axis of the graph. Look on the vertical axis for the number 10, and then look at which state's bar aligns with 10. The state is Alabama.

2. **(3) Mississippi** (Application) The chances of avoiding a hurricane are probably best in the state that had the fewest hurricanes during the 96 years covered by the graph. Of the states shown, Mississippi had the fewest hurricanes—only 8.

3. **(2) photosynthesis** (Application) The energy released by the fusion of hydrogen is absorbed in the form of sunlight by green plants. The plants turn the energy into food, and all living things on Earth depend on this process for their own energy.

4. **(1) precipitation** (Comprehension) According to the diagram, water moves from the atmosphere to Earth's surface by condensing and falling as precipitation.

5. **(2) drinking liquids, excreting urine, and exhaling water vapor** (Application) Humans contribute to the water cycle by drinking fresh water and other liquids that contain fresh water. They return water to the surface by urinating. They return water to the atmosphere by exhaling water vapor, a waste product of cellular respiration.

6. **(3) nuclear power** (Comprehension) Look for the energy source that occupies about one-fourth of the circle graph. That source is nuclear power, which provides 27.4% of Minnesota's electricity.

PROGRAM 18
CHEMISTRY

Practice 1 (page 61)

1. **space**

2. **weighing** Weight is a measure of mass.

3. **gravity** Weight is a measure of the pull of gravity on an object. The pull of gravity is weaker on the moon than it is on Earth. So objects weigh less on the moon than they do on Earth.

4. **properties** The properties, or characteristics, of matter can be seen, touched, tasted, and smelled.

5. **Density**

6. **denser** Even though the volume of the steel and the air are the same, there is much more matter in 1 cubic foot of steel than in 1 cubic foot of air.

7. **protons, neutrons, electrons** Answers can be given in any order as long as all three types of particles are mentioned.

8. **neutral** The positive and negative charges cancel one another out.

9. Answers will vary. Sample: Plastic comb— smooth texture, black, medium hardness, not very dense, somewhat flexible, odorless, and tasteless.

Practice 2 (page 63)

1. c. gas

2. a. solid

3. b. liquid

4. a. melting and boiling   In melting, heat is added to a solid, and it becomes a liquid. In boiling, heat is added to a liquid, and it becomes a gas.

5. c. condensing and freezing   In condensing, heat is removed from a gas, and it becomes a liquid. In freezing, heat is removed from a liquid, and it becomes a solid.

6. Melting and freezing both involve solids and liquids, but they are opposite processes. In melting, heat is added to a solid, and it becomes a liquid. In freezing, heat is removed from a liquid, and it becomes a solid.

7. Answers will vary. Sample answer: A puddle of rain drying up in the sun is a common example of evaporation.

Practice 3 (page 65)

1. True

2. False   More than 110 elements have been discovered, but not all of them have been found to occur naturally on Earth.

3. True   All elements with atomic numbers of 93 and above have been made in laboratories.

4. True

5. False   Some elements are reactive, and other elements are unreactive. The unreactive elements do not react easily with other elements.

6.

11
Sodium
Na

7.

17
Chlorine
Cl

Practice 4 (page 67)

1. True

2. True

3. True

4. False   The chemical formula for water is H_2O.

5. False   In a mixture, the substances can be combined in any proportion.

6. False   The components of mixtures can be separated through a variety of different physical methods, depending on the components of the mixture. For example, a mixture of iron filings and sawdust can be separated using a magnet.

7. Both compounds and mixtures are combinations of substances. Compounds are chemical combinations of two or more elements in a definite proportion. Mixtures are physical combinations of two of more substances, put together in any proportion.

Practice 5 (page 69)

1. f. exothermic reaction

2. d. products

3. e. reactants

4. a. chemical reaction

5. b. endothermic reaction

6. c. activation energy

7. The mass of the reactants equals the mass of the products in a chemical reaction. Substances change but they do not gain or lose mass in total.

8. In exothermic reactions, energy is released; in endothermic reactions, energy is absorbed.

9. The chart indicates that photosynthesis is an endothermic reaction. This means that energy is absorbed, generally as sunlight.

Practice 6 (page 71)

1. carbon dioxide

2. baking powder

3. acid

4. yeast

5. water, oxygen (in either order)

6. grease, water

7. Unleavened breads do not contain a leavening agent such as yeast or baking powder.

8. Hydrogen peroxide doesn't foam unless catalase is present. Catalase is in blood, not unbroken skin. Therefore, when you pour hydrogen peroxide on unbroken skin, nothing happens.

9. Soaps are compounds made of an alkali metal and a fatty acid. Detergents are compounds made of the sodium salts of acids that contain sulfur.

GED Thinking Skill Practice: Analysis (page 73)

(5) help prevent some cancers Marinating meat and fish before grilling seems to lower the production of heterocyclic amines produced in the food when it is grilled. These compounds have been linked to breast and colon cancer in laboratory animals. Therefore, eating less of them may have a health benefit of helping to prevent cancer in humans.

GED Graphic Skill Practice: Using Tables (page 75)

1. (2) energy level 2 (Comprehension) To answer this question, locate the "Maximum Number of Electrons" column, and then go down to the row that has 8 electrons. Read across to the left to see what energy level holds a maximum of 8 electrons. The answer is energy level 2.

2. (3) nitrogen (Analysis) To answer this question, you must compare the boiling points listed in the second column. Look for the lowest boiling point, which would be one of the two that has a negative value. The lowest is −196°C, the boiling point of nitrogen.

GED Review (pages 76–77)

1. (3) 1 teaspoon (Comprehension) First look in the "Use" column for feeding houseplants every four weeks. Then read across that row to the amount of plant food that should be dissolved in 1 gallon of water. The amount is 1 teaspoon. Note that the plant food is the solute and the water is the solvent in this solution.

2. (3) houseplants fed with every watering (Analysis) The weakest solution is the one with the least amount of solute per volume of solvent. In the table, the amount of solvent, the water, remains the same— 1 gallon. So you must look at the amounts of solute, or plant food. The smallest amount is $\frac{1}{4}$ teaspoon, which is for houseplants fed with every watering.

3. (4) Both consist of carbon, hydrogen, and oxygen atoms. (Analysis) Choices (1), (2), and (3) are incorrect because each is true of either sucrose or glucose, but not both. Choice (5) is incorrect because only sucrose is white table sugar.

4. (1) It weakens the strength of the cleaner. (Analysis) Mixing the cleaner with water or milk causes the cleaner to become less concentrated, or weaker.

5. (4) a chemical reaction producing a poisonous gas (Application) Since two liquids, ammonia and bleach, are combining without added heat to produce a dangerous gas, a substance with completely different properties, this must be a chemical reaction.

6. (1) adding heat (Analysis) According to the passage, in most cases adding heat to a solid causes it to melt. In sublimation, adding heat causes a solid to change to a gas.

PROGRAM 19
PHYSICS

Practice 1 (page 81)

1. force

2. **Gravity**   The huge mass of Earth gives it great gravity.

3. **Velocity**   Remember, velocity has two elements: speed and direction.

4. **second law of motion**

5. **force**

6. **equal, opposite** (either order)   This law is sometimes stated as "For every action, there is an equal and opposite reaction."

7. **the first law of motion**   Unless the Voyager spacecraft is acted upon by a force, such as the gravity of a comet, asteroid, or someday another star, it will continue in its uniform motion away from our solar system.

Practice 2 (page 83)

1. **c. energy**

2. **d. potential energy**

3. **f. kinetic energy**

4. **e. effort force**

5. **a. machine**

6. **b. work**

7. Both the Law of Conservation of Matter and the Law of Conservation of Energy state that something can be neither created nor destroyed but simply changed in form. In one law, the "something" is matter; in the other law, it is energy.

8. Answers will vary. Possible answers:

   a. inclined plane: ramp, screw

   b. pulley: block and tackle, crane

   c. lever: crowbar, seesaw, scissors

   d. wedge: doorstop, knife blade, needle

9. In the diagram of potential energy, the roller-coaster car should be at the top of the hill. In the diagram of kinetic energy, it should be moving up or down one of the slopes.

Practice 3 (page 85)

1. **True**   The waves move the cork up and down, but they don't push the cork to shore. (That is done by the wind.)

2. **True**

3. **True**

4. **False**   The distance between one crest and the next is a wavelength.

5. **False**   The height from the midpoint to the top of the crest (or from the midpoint to the bottom of the trough) is the amplitude.

6. A beach ball or a chip of wood bobs up and down on a water wave because the wave is transmitting energy through the water, not moving the water forward.

7. Since sound travels about 1 mile in 5 seconds, divide 10 seconds by 5. The lightning strike is about 2 miles away.

8. Seismic waves travel through solid ground.

Practice 4 (page 87)

1. electromagnetic

2. electric, magnetic (either order)

3. space   Unlike sound waves or water waves, electromagnetic waves do not need to travel through matter.

4. electromagnetic spectrum

5. c. microwaves

6. a. radio waves

7. e. visible light

8. e. visible light

9. h. gamma rays

10. b. radar waves

11. d. infrared waves

12. g. X rays

13. f. television waves

14. Sound waves must travel through a medium such as air or water. Electromagnetic waves can travel through a medium or through the vacuum of space.

15. 1 cm   You can find this information on the diagram of the electromagnetic spectrum, below the band showing microwaves.

Practice 5 (page 89)

1. True

2. False   The atoms are lined up in a bar magnet.

3. True

4. False   The magnetic field is strongest near the poles of a magnet.

5. True

6. Answers will vary. Sample answers: refrigerator magnets, videotapes, telephone receivers, loudspeakers, computers.

7. Turning off an electromagnet enables you to stop its action. For example, if you are lifting scrap metal, you can put the metal down by turning off the electromagnet. The metal will fall.

Practice 6 (page 91)

1. a. electrons

2. d. magnetic field

3. a. battery

4. b. circuit

5. c. start and stop the current

6. d. the light bulb

7. The flashlight is off. That's because the switch is in the off position, causing a break in the circuit. The current doesn't flow because the circuit is incomplete.

8. Answers will vary. Possible answers: At home, lights, power for appliances and electronic devices like TVs. At work, lights, motors in factory machines, computers.

GED Thinking Skill Practice:
Evaluation (page 93)
(4) Nearsightedness can be corrected by
eyeglass lenses. The passage explains that
the eyes of nearsighted people are too long,
so the images are focused in the wrong place in
the eyeballs. With a corrective lens, as shown
in the diagram, the image is focused on the
retina. With glasses, nearsighted people can see
as clearly as do people with normal vision.

GED Graphic Skill Practice:
Using Diagrams (page 95)
(1) a barrier through which light rays
cannot pass (Comprehension) The diagram
shows that when the light ray reaches the surface
of the mirror, it is stopped and reflected. It then
travels away from the mirror. This suggests that
the mirror is something through which light
cannot pass.

GED Review (pages 96–97)
1. **(4) reducing friction** (Analysis) The passage
explains that the roughness of surfaces causes
friction when they slide past one another. In
the diagram, you can see that the lubricant
works by separating the two surfaces. Now
when they slide past one another, they do not
touch, which decreases the friction.

2. **(1) preventing wear on machine parts**
(Application) Machines that are not
lubricated wear out quickly as the metal parts
grind against one another. For example, that's
why oil is used to lubricate the moving parts
of car engines.

3. **(5) swing the cork in circles and then let go**
(Evaluation) Demonstrating centripetal force
involves showing what happens when the
force is applied (the object moves in a circle)
and then when it is not applied (letting go of
the string). Choices (1), (2), and (3) do not
involve any circular motion, so centripetal

force is not applied at all. Choice (4) involves a component of circular motion with the swinging pendulum but does not include letting the pendulum go, so it does not fully demonstrate centripetal force.

4. **(3) conversation** (Comprehension)   Look for 50–60 in the column labeled "Decibels." Then read across the column to see what type of sound is generally that loud.

5. **(4) airport ground workers** (Application) Since people regularly exposed to decibel levels higher than 80 are at risk of damaging their hearing, look in the table for sounds greater than 80 decibels. Of the sounds listed, one is jet aircraft. Airport ground workers would be exposed to noise from jet engines on a regular basis, so if they didn't wear ear protection, they would be at high risk for hearing loss. None of the other occupations requires exposure to higher noise levels on a regular basis.

6. **(2) They reflect different color light waves.** (Comprehension)   According to the passage, the object a color appears to be depends on what color light waves it absorbs and what color light waves it reflects.

7. **(3) Clothing appears to be different colors in fluorescent light and sunlight.** (Evaluation) The properties of the clothing remain the same, so the fact that it appears a different color under different light conditions must mean that the light conditions are different. In this case, fluorescent light and sunlight are slightly different combinations of color light waves, although they appear to us to be the same. To answer this question correctly, you must look for the facts that support the conclusion. Choices (1), (2), and (5) are all true, yet they are incorrect because they do not support the conclusion. Choice (4) is false and therefore incorrect.

Science Resources

DNA

Inside the nucleus of cells are chromosomes, microscopic threads of genetic material. These chromosomes have all the information needed to regulate the cell's activities and to reproduce. The chromosomes consist of DNA coiled around a protein core. If the DNA in a cell were uncoiled, it would be about 13 feet long. A gene is a short length of DNA that codes for a particular protein.

The DNA molecule itself has a double helix (twisted ladder) structure. The "rungs" of the ladder are made of base pairs. There are four bases in DNA: adenine, thymine, cytosine, and guanine. Adenine always pairs with thymine, and cytosine always pairs with guanine. A sequence of three base pairs codes for a particular amino acid; amino acids are the building blocks of proteins.

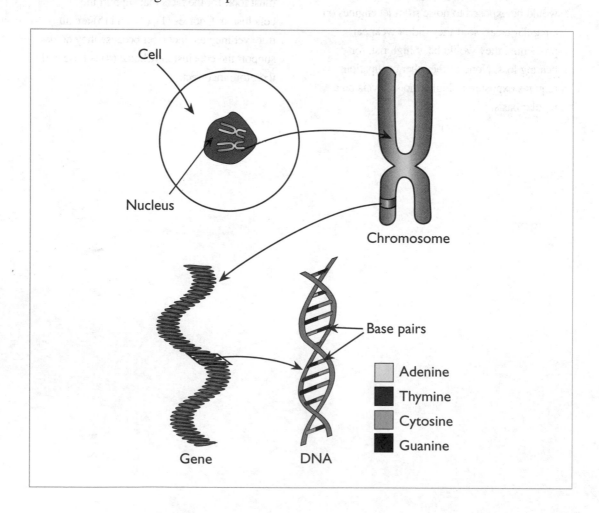

The Solar System

For a description of the inner and outer planets, see pages 48–49. The following table summarizes some important data about the planets.

Planet	Distance from Sun (km)	Diameter at Equator (km)	Rotation Period (in Earth days)	Orbit Period (in Earth years)	Number of Moons
Mercury	57,909,175	4,879	58.6	0.2	0
Venus	108,208,930	12,104	243.0	0.6	0
Earth	149,597,890	12,756	1.0	1.0	1
Mars	227,936,640	6,786	1.0	1.9	2
Jupiter	778,412,020	142,984	0.4	11.9	39
Saturn	1,426,725,400	120,536	0.4	29.5	30
Uranus	2,870,972,200	51,118	0.7	84.0	20
Neptune	4,498,252,900	49,528	0.7	164.8	8
Pluto	5,906,376,200	2,290	6.4	247.9	1

PLANET FACTS

Space exploration began less than fifty years ago with the Soviet Union's Sputnik 1, the first artificial satellite. Major space missions are shown below.

Mission	Year	Crew?	Accomplishment
Sputnik 1 (Soviet Union)	1957	No	First artificial satellite to orbit Earth
Mercury	1961–1963	Yes	First U.S. astronaut in orbit around Earth
Apollo 11	1969	Yes	First men land on moon
Pioneer 10 and 11	1972–1973	No	First space probes to Jupiter and Saturn
Voyager 1 and 2	1977–1989	No	Flew by and collected data on Jupiter, Saturn, Uranus, Neptune and their moons
Space Shuttle	1981–present	Yes	Reuseable spacecraft; launching and servicing satellites; scientific experiments
Mir (Soviet Union)	1986–2001	Yes	Space station; long-term crew missions; scientific experiments
Galileo	1995–1997	No	Flew by and collected data on Jupiter and its moons
Pathfinder	1997	No	Robotic exploration of Mars' surface
International Space Station	2000–present	Yes	Space station; long-term crew missions; scientific experiments

MAJOR SPACE MISSIONS

Periodic Table of the Elements

In the periodic table, chemists arrange all the elements in a grid. They are arranged in order of atomic number and the way their electrons are organized. (The atomic number is the number of protons in the nucleus of one atom of the element.) In the periodic table, the horizontal rows are called **periods.** The vertical columns are called **groups** or **families.**

In each group, each element has the same number of electrons in its outer electron shell, giving it chemical properties similar to those of the other elements in the same group. For example, all the elements in Group 1 have one electron in the outer electron shells of their atoms. Because of this, elements in Group 1 have similar chemical properties. They all react easily with other elements. In contrast, the elements in Group 18 have full outer electron shells. These elements, called the noble gases, almost never react with other elements.

As you go down the table, the periods get longer because the atoms have more electron shells, so there are more elements in each period. Periods 6 and 7 are 32 elements long. That makes the table so wide that a series of elements from each of these two periods, the lanthanides and the actinides, are usually shown below the table.

THE PERIODIC TABLE

1 Hydrogen H																	2 Helium He
3 Lithium Li	4 Beryllium Be											5 Boron B	6 Carbon C	7 Nitrogen N	8 Oxygen O	9 Fluorine F	10 Neon Ne
11 Sodium Na	12 Magnesium Mg											13 Aluminium Al	14 Silicon Si	15 Phosphorus P	16 Sulfur S	17 Chlorine Cl	18 Argon Ar
19 Potassium K	20 Calcium Ca	21 Scandium Sc	22 Titanium Ti	23 Vanadium V	24 Chromium Cr	25 Manganese Mn	26 Iron Fe	27 Cobalt Co	28 Nickel Ni	29 Copper Cu	30 Zinc Zn	31 Gallium Ga	32 Germanium Ge	33 Arsenic As	34 Selenium Se	35 Bromine Br	36 Krypton Kr
37 Rubidium Rb	38 Strontium Sr	39 Yttrium Y	40 Zirconium Zr	41 Niobium Nb	42 Molybdenum Mo	43 Technetium Tc	44 Ruthenium Ru	45 Rhodium Rh	46 Palladium Pd	47 Silver Ag	48 Cadmium Cd	49 Indium In	50 Tin Sn	51 Antimony Sb	52 Tellurium Te	53 Iodine I	54 Xenon Xe
55 Caesium Cs	56 Barium Ba	57 – 70 *	71 Lutetium Lu	72 Hafnium Hf	73 Tantalum Ta	74 Tungsten W	75 Rhenium Re	76 Osmium Os	77 Iridium Ir	78 Platinum Pt	79 Gold Au	80 Mercury Hg	81 Thallium Tl	82 Lead Pb	83 Bismuth Bi	84 Polonium Po	85 Astatine At	86 Radon Rn
87 Francium Fr	88 Radium Ra	89 – 102 **	103 Lawrencium Lr	104 Rutherfordium Rf	105 Dubnium Db	106 Seaborgium Sg	107 Bohrium Bh	108 Hassium Hs	109 Meitnerium Mt	110 Ununnilium Uun	111 Unununium Uuu	112 Ununbium Uub		114 Ununquadium Uuq				

*Lanthanides

57 Lanthanum La	58 Cerium Ce	59 Praseodymium Pr	60 Neodymium Nd	61 Promethium Pm	62 Samarium Sm	63 Europium Eu	64 Gadolinium Gd	65 Terbium Tb	66 Dysprosium Dy	67 Holmium Ho	68 Erbium Er	69 Thulium Tm	70 Ytterbium Yb

**Actinides

89 Actinium Ac	90 Thorium Th	91 Protactinium Pa	92 Uranium U	93 Neptunium Np	94 Plutonium Pu	95 Americium Am	96 Curium Cm	97 Berkelium Bk	98 Californium Cf	99 Einsteinium Es	100 Fermium Fm	101 Mendelevium Md	102 Nobelium No

Alkali Metals

Transition Metals

Nonmetals

The Scientific Method

The scientific method is actually a set of processes that scientists use to add to the body of scientific knowledge. It can vary depending on the material or subject being studied.

To illustrate the scientific method, we'll use an experiment performed in 1668 by Italian physician Francisco Redi. At the time, people thought that living organisms could arise from nonliving things. For example, they thought that the rats in cities arose from the sewage flowing in the streets. They thought that frogs came from mud. The idea that living things arise from nonliving things is called spontaneous generation.

Like others, Redi observed that maggots grow on rotting meat and flies swarm around it. Redi performed an experiment to prove that the maggots come from flies, rather than from meat. (Maggots are the larva stage of the fly's life cycle.)

Step 1. Identify a problem

Redi's problem was to figure out where maggots on rotting meat come from.

Step 2. Collect information

Redi observed that maggots grow on rotting meat. He also saw flies swarming on meat.

Step 3. Form a hypothesis based on the information gathered

Redi hypothesized that the maggots came from the flies, rather than from the meat.

Step 4. Test the hypothesis

Redi set up three groups of wide-mouthed jars containing meat.

- The first group was the control group. The jars were open so the meat would be exposed to whatever it was exposed to normally.
- The second and third groups were the variables. The second group of jars was covered with netting. Air could circulate, but flies couldn't get in. The third group of jars was sealed completely. Nothing could get in.

Redi then recorded the presence or absence of flies and maggots. In the group of open jars, he saw flies and then maggots on the meat. In the jars covered with netting, he observed flies on the netting but not in the jars. He later saw a few maggots on the meat in those jars. In the completely sealed jars, he did not see any flies or maggots.

Step 5. Draw conclusions based on the data that was collected

Redi concluded that the maggots came from the flies, not the meat. He based this conclusion on several observations: First, flies could enter the uncovered jars and lay eggs on the meat. These later grew into maggots. Second, flies laid eggs on the netting; the eggs fell through and developed into maggots on the meat. Third, no flies, eggs, or maggots could enter the sealed jars, so they were not present in those jars. Redi's experiment gave strong evidence refuting the idea that maggots come from rotting meat.

National Science Education Standards

GED Science Test items draw on topics from life science, earth science, chemistry, and physics. They also draw on some basic themes in science.

Unifying Concepts and Processes in Science

It's easy to think of science as a collection of individual facts, like the boiling point of water or the life cycle of a frog. However, science is really an interrelated set of processes and concepts. These cross the boundaries between the life and physical sciences. For example, the process of accurate measurement is important whether you are measuring blood pressure or the mass of chemical compounds. The concept of organization into systems applies to body organs as well as to the solar system.

Science as Inquiry

All scientists investigate the world. They ask questions, plan and carry out experiments, gather data, think critically about what they observe, and draw conclusions about the relationship of evidence and explanations. Redi's experiment to determine where maggots come from is a good example of science as inquiry (see page 127). The thinking skills associated with scientific inquiry can be applied to any area of science.

Science and Technology

The knowledge gained through scientific research is often used in technology. For example, knowledge about DNA is used to genetically engineer drugs and foods. However, technology also influences scientific research and thinking. For example, the orbiting Hubble telescope, able to "see" farther into space than any previous telescope, has led scientists to formulate new theories about the nature of the universe.

Science in Personal and Social Perspectives

Science is extremely important in modern life. Advances in science affect our personal and community health. Scientific knowledge influences population growth, the use of natural resources, and environmental quality. It influences our reactions to nature's changes on a global scale (global warming) as well as on a local scale (soil erosion). However, people don't take an exclusively scientific approach to these issues. Politics, economics, and culture also play a big role.

History and Nature of Science

Science is ultimately a human effort. It is conducted by people working alone and together, in different cultures, and at different periods of time. All these factors influence the accumulation of scientific knowledge and the role of science in society.